T0205522

Springer Tracts on Transportation and Traffic

Volume 18

About this Series

The book series "Springer Tracts on Transportation and Traffic" (STTT) publishes current and historical insights and new developments in the fields of Transportation and Traffic research. The intent is to cover all the technical contents, applications, and multidisciplinary aspects of Transportation and Traffic, as well as the methodologies behind them. The objective of the book series is to publish monographs, handbooks, selected contributions from specialized conferences and workshops, and textbooks, rapidly and informally but with a high quality. The STTT book series is intended to cover both the state-of-the-art and recent developments, hence leading to deeper insight and understanding in Transportation and Traffic Engineering. The series provides valuable references for researchers, engineering practitioners, graduate students and communicates new findings to a large interdisciplinary audience.

Indexed by SCOPUS, WTI Frankfurt eG, zbMATH, SCImago.

More information about this series at http://www.springer.com/series/11059

Konstantinos Tzanakakis

Managing Risks in the Railway System

A Practice-Oriented Guide

Konstantinos Tzanakakis
Athens, Greece

ISSN 2194-8119 ISSN 2194-8127 (electronic)
Springer Tracts on Transportation and Traffic
ISBN 978-3-030-66268-4 ISBN 978-3-030-66266-0 (eBook)
https://doi.org/10.1007/978-3-030-66266-0

This Springer imprint is published by the registered company Springer Nature Switzerland AG
The registered company address is: Gewerbestrasse 11, 6330 Cham, Switzerland

Take calculated risks. That is quite different from being rash.

George S. Patton

Preface

In the past 30–50 years, our life has changed fundamentally. The most significant difference is the level of complexity in how we manage things, our business, our life. There have always been complex systems. Complexity, however, affects almost everything that we touch: the technology that we use, how we do our business, the jobs that we do every day, etc.

Our organizations are also more complicated than they were in the past! Nowadays, organizations have many (and sometimes complicated) processes, many people are involved in these processes, many rules have been established for our organizations that manage the railway system, etc.

All activities performed by an organization involve risk. We now have systems of systems, and this affects the railway system's safety. **Controlling safety is about managing the risks**. Over the past decades, risk management has become a more and more critical activity within organizations. This became an integral part of good management practice.

This book is intended to help organizations to elaborate and implement an effective framework for risk management, so they can take decisions and manage their risks. It is a guide for risk management, so to enable public, private companies, groups, and individuals to manage risks.

The book is structured in five chapters as follows:

Chapter 1: An Introduction to Risk Management/Setting the Scene

Risk is a common part of doing business within our organizations, but risk can be controlled. This chapter introduces risk management (as an integral part of good management practice) and presents the history and evolution of risk management. It explains how risk management within organizations is becoming an increasingly important activity. Many examples clarify the basic terms of risk management.

The rail system is presented as a complex system. Finally, the system life cycle as an arrangement of phases from initial concept through to decommissioning and disposal will be discussed.

Chapter 2: The Concept of Risk Management

This chapter presents the concept of risk management as a critical component of the strategic management of all organizations. The necessity of risk management is discussed: The alternative to risk management is risky management.

Current frameworks for risk management around the world are provided, and different approaches are presented: in the civil aviation sector, as per the industry approach, the Enterprise Risk Management, the Project Management Institute, the Intergovernmental Organization for International Carriage by Rail (OTIF), and the risk management framework in different countries: the UK, the European Union, the USA, Japan, Canada and Australia/New Zealand. Then, a comparison of previously mentioned risk management frameworks is provided.

The benefits and principles of an effective and robust risk framework are presented.

Finally, the way to manage changes in our system and the process of managing them is discussed.

Chapter 3: The Process of Risk Management

This chapter is dedicated to the processes of risk management. The key elements of the risk management process are discussed in detail (setting the context, identifying risks, analysing risks, risk assessment, treating risks, monitoring and review, and finally communicating and consulting).

Chapter 4: Risk Assessment Techniques

This chapter introduces risk assessment techniques and their applicability (for risk identification, risk analysis and risk evaluation). It also discusses the risk management plan.

Chapter 5: Health Risk Management (For Staff with Safety-Critical Positions)

This chapter provides practical guidance to assist rail organizations in performing health risk assessments for staff with safety-critical positions.

The primary audience for this book is engineers from the railway industry (civil, electrical, mechanical, signalling and telecom engineers, safety engineers) working at the ministries of transport, government authorities, railway organizations, public and private railway developers (infrastructure managers, train operating companies, metros), consultants and contractors, safety assessors, etc.

After reading this book, the reader will:

- be familiar with the terms related to risk and safety
- get to know a roadmap for risk management and
- be able to understand risk management techniques,

so, effectively manage the risks of the railway system.

Athens, Greece
February 2020

Konstantinos Tzanakakis

Contents

About the Author

Konstantinos Tzanakakis received Civil Engineering Master's Degree from the University of Hannover, Germany, and a Master Executive MBA from the Athens University of Economics and Business, Greece. Out of his 35-year experience in the railway sector, he was serving as Director of the Greek Railways Organization for eight years. He worked in Serbia, being involved in the restructuring of the Serbian Railways between 2011 and 2013. Between 2015 and 2019, he served as Senior Railway Expert at the Ministry of Transport and Communications in Oman, taking care of various railway institutional issues, railway policies and the establishment of a Railway Authority in Oman. Between 2013 and 2014, he made a major contribution to the development of the Oman National Railway Network, while working for the Oman Rail Company. Currently, he is working as Project Manager for two railway projects in both Albania and Egypt.

He is also Founder of RAILHOW (https://railhow.com/), an initiative with the purpose of developing and delivering well-researched and practical-oriented engineering and leadership digital resources and training. With a strong professional focus on the efficiency of the railway system, he is also Author of the book *The Railway Track and its Long-Term Behaviour*, published by Springer in 2013. In recent years, he has been invited as a speaker to several railway congresses; he held various workshops about best practices in track maintenance and strategies for keeping a high-performance railway track.

Connect with Me:

LinkedIn: Kostas Tzanakakis

Author webpage: www.railhow.com

Abbreviations

ALARA	As Low As Reasonably Achievable
ALARP	As Low As Reasonably Possible
CBA	Cost-Benefit Analysis
CSI	Common Safety Indicator
CSM	Common Safety Method
CSM RA	Common Safety Method (CSM) for Risk Evaluation and Assessment
CST	Common Safety Target
DPA	Death Per Annum
EDPA	Equivalent Deaths Per Annum
EFAR	Equivalent Fatal Accident Rate
ES	European Standard
EU	European Union
FAR	Fatality Accident Rate
FHA	Functional Hazard Analysis
FMEA	Failure Mode and Effects Analysis
FMECA	Failure Mode, Effects and Criticality Analysis
FTA	Fault Tree Analysis
HAZOP	Hazard and Operability Studies
ISO	International Organization for Standardization
ORR	Office of Rail and Road (UK)
PMI	Project Management Institute (https://www.pmi.org)
RAMS	Reliability, Availability, Maintainability and Safety
SMS	Safety Management System
UK	United Kingdom

List of Figures

List of Tables

Chapter 1
An Introduction to Risk Management/ Setting the Scene

> The worst thing is to rush into action before the consequences have been properly debated ….. We (Athenians) are capable at the same time of taking risks of estimating them beforehand.[1] (Pericles 430 *v.* Chr.)
>
> The biggest risk is not taking any risk… In a world that is changing really quickly, the only strategy that is guaranteed to fail is not taking risks. (Mark Zuckerberg)
>
> A ship is safe in harbour, but that's not what ships are for. (John A. Shedd)

1.1 Risk Is Part of Everyday Life

Risk is part of daily life since we are on earth. In our day to day life, there are many circumstances, where we have to take risks, which involves exposure to lose or danger. Risk surrounds us. Risk is present in every aspect of our lives. The term "risk" means the chance that **something "bad" may happen**. **Risk describes a situation with a possibility of loss or danger**. Figure 1.1 presents situations of our daily life where risks resulted in "bad" situations.

Risk is common in doing business in our Organizations but risk can be managed; in some cases, the risks may be so small that it is not worth the effort to manage them; in other cases, risks connected with the activities of an Organization need to be carefully followed and managed.

[1]This statement recorded by Thucydides (460–400 B.C.) was made by Pericles at the public ceremonial burial of the first Athenians killed in the war with the Spartans. Here Pericles describes the virtues of democracy as practiced by Athens.

K. Tzanakakis, *Managing Risks in the Railway System*, Springer Tracts on Transportation and Traffic 18, https://doi.org/10.1007/978-3-030-66266-0_1

Fig. 1.1 Risks are surrounding us

There are different risk types, i.e.:

A. As per the "nature" of the risk:
1. Financial Risks: Financial risk is usually linked to insurance.
2. Technological Risks, related i.e. to data security and the security of records.
3. Physical and environmental risks within areas such as the construction of assets, their maintenance and operation, as it is the railway infrastructure and the operation of trains (this book focuses in this type of risks).

B. As per the "scope" of risk management:
1. Enterprise Risk Management (ERM)/Risk Management for Organizations
2. Project Risk Management, that refers to managing the risks of a project.

Risk management within Organizations is becoming an increasingly important activity. It is an integral part of good management practice and a crucial element of corporate governance, that is the system of rules, practices and processes by which an Organization is directed and controlled. We call "risk management" the tools, techniques, processes and methodologies that have been developed in order to improve the ability of our Organizations to deal with the uncertainty, and especially with its negative impact.

Risk management should be incorporated in an Organization's culture, embedded into the Organization's philosophy, practices and business processes, so that everyone within the Organization is involved in risk management.

As we discussed, risks strictly mean "bad news"; however, some risks can be positive. Negative risks are unwanted and potentially can cause serious problems; positive risks, on the other hand, are opportunities and are desired by our Organizations and the Stakeholders and may positively affect the performance of our Organizations. In this book, where "risk" is mentioned, to be understood "negative risks".

This book is focussing on the physical and environmental risks of a (Railway) Organization and not with managing the risks of projects.

1.2 History and Evolution of Risk Management

The term risk may be traced back to classical Greek "ρίζα", meaning root, later used in Latin for the cliff, due to the hazards of sailing along rocky coasts.

Some historians believe that the concept of risk arose from gaming [1].

The Code of Hammurabi was the first known Risk Code of Practice. It is a Babylonian code of ancient Mesopotamia dating back to about 1754 B.C. Next, some precepts of interest[2]:

- Precept nr. 229: *If a builder builds a house for someone, and does not construct it properly, and the house which he built fall in and kill its owner, then that builder shall be put to death.*
- Precept nr. 230: *If it kills the son of the owner, the son of that builder shall be put to death.*
- Precept nr. 230: *If it kills a slave of the owner, then he shall pay slave for slave to the owner of the house.*
- Precept nr. 232: *If it ruins goods, he shall make compensation for all that has been ruined, and since he did not construct properly this house which he built, and it fell, he shall re-erect the house from his own means.*

In order to follow the Code of Hammurabi, risk management should be applied.

Also in ancient Greece, several thousands of years ago, the role of chance in life and its outcomes was explicitly recognised in Greek mythology. The three fates, Cloth, Lachesis and Atropos determined the outcomes for all human life, the probability of future fame and fortune, and of life and death itself. *Clotho* was forever spinning and creating the thread of life; *Lachesis* was casting the lots that determine human's destiny in the woven fabric of life, and blind *Atropos* cut with her sharp shears then the thread to end that life.

[2]For specifics of the Code see: http://avalon.law.yale.edu/ancient/hamframe.asp.

In recent times, in the last decades, two disasters have occurred, which we will be referring to next. Because of those disasters but also because of other disasters, risk management has become a necessary management activity.

On the night of 2–3 December of 1984, there was a gas leak incident at the pesticide plant of Union Carbide India Ltd. in Bhopal, India. The **Bhopal disaster** (also referred to as the Bhopal gas tragedy), was considered to be the world's worst industrial disaster (Fig. 1.2).

Another disaster was the **Exxon Valdez oil spill** in 1989. The Exxon Valdez oil spill occurred in Alaska on March 24, 1989, when Exxon Valdez, an oil tanker struck a reef. 10.8 million US gallons (or a mass of 35,000 metric tonnes) of crude oil spilt over the next few days. It is considered to be one of the most devastating human-caused environmental disasters.

The risk management study began after the Second World War. Risk management has long been linked to the use of market insurance to protect individuals and businesses from various accident-related losses. Table 1.1 presents the main phases of the evolution of risk management after 1950.

1.3 The Risks of Our Complex World

Live was simple in the past.....

Let us discuss the risks of our complex world on some examples.

Example 1: The First Car in History That Got a Fine for Speeding…
On the 28/1/1896, Walter Arnold was driving in Kent (UK) a "Benz Motor Carriage", the first car in history that got a fine for speeding. The speed limit was 3 km/h. The owner and driver -as mentioned in the police report- was running with a speed of 12 km/h, that is 4 times the speed allowed. Shortly after this event, the limits have been increased to 22 km/h (Fig. 1.3).

Example 2: The First Fatal Traffic Accident in Greece…
The first fatal traffic accident in Greece occurred in Athens, on 4/3/1907, when there were only seven cars. The Commander of the Police, the next day, called the seven drivers in Athens and made very strict recommendations to drive with caution. He made it clear that "within the city, no more than 10 km of speed may be allowed". The newspapers of this era had raised this issue to the first issue, saying "*Seven cars are circulating, and we are mourning victims ... Imagine what happens if they become seventy*"!

Nowadays: Living with Complexity
Imagine turning the clock back 50–70 years. How did things happen in 1950? Compared to 50–70 years ago, how different is your working life today? The most profound difference is the level of complexity that we have to manage things, to manage our business, our life.

Fig. 1.2 The Bhopal disaster (left), Exxon Valdez oil spill (right)

Table 1.1 The evolution of risk management

Risk management		
In the past: compliance	Currently: prioritise problems	In the future: business optimisation
• Identify problems • Rank them • Demonstrate that each risk has a control (generally a standard procedure) • Monitor controls	• Identify problems • Rank them • Check whether the risk level is above the target level (qualitative) • Implement better controls, beginning with the highest risks • Monitor implementation	• Identify any possible problems and opportunities • Understanding the causes and factors that affect likelihood and consequences • Optimise treatment considering ◦ Effectiveness of current and proposed controls ◦ Causal factors ◦ Costs and benefits of treating the risk ◦ Costs and benefits of taking the risk • Treat according to risk appetite • Monitor and feedback

There were always complex systems. Complexity, however, has gone from something found mainly in large systems, like cities, to something that affects almost everything we touch: the technologies we use, the way we do our business, the jobs we do every day, and so on.

Systems used to be simple… a single person could fully understand them… but no longer! Systems that used to be separate are now interconnected and interdependent, meaning they are more complex, by definition.

THE QUESTION OF HORSELESS CARRIAGES.—At the Tunbridge Police-court, Mr. Walter Arnold, the owner of a horseless carriage, was summoned on four informations with reference to using a horseless carriage on the highway. The first was for using a locomotive without a horse from the County Council, the second for having less than three persons in charge of the same, the third for going at a greater rate than two miles an hour, and the fourth for not having his name and address placed on the machine.—The evidence was that the carriage was going at the rate of eight miles an hour.—Mr. Cripps, who defended, contended that the machine was not one contemplated when the Locomotive Acts were passed, and said that in the past these carriages had been used by Sir David Salmons and the Hon. Evelyn Ellis without any notice being taken. If the Bench considered that the carriage was a locomotive, he asked for the imposition of a nominal penalty.—The Bench inflicted a penalty of 5s. and 2*l*. 0s. 11d. costs for using the carriage without a locomotive horse, and 1s. and 9s. costs in each of the other three cases—4*l*. 7s. altogether.

Fig. 1.3 The first car in history that got a fine for speeding and the news at "London Daily News"

Organizations are also more complex than they were in the past! Nowadays Organizations have many (and sometimes complicated) procedures, and more people are involved in these processes, several rules for managing the rail system and its parts during their life cycle have been put in place, strategies have been developed that must be implemented, etc.

Is complexity impacting the safety of our systems? Does complexity affect our systems' efficiency, and if so, how? Let us look at a few examples from the transport sector.

- **Automobile industry**: The cars in the past have been very simple. Cars today are too complicated with system of systems. Autonomous driving is already a reality. Does this affect the safety of driving?
- **Aviation sector**: An *aircraft* is a complex system. Aeroplanes are equipped with many control systems, digital computers and associated accessories for controlling all parts of an aeroplane. Does this affect the safety of flying?
  Airports are complex systems. They are characterised by complex interdependencies between different systems, e.g. check-in, security and retail areas and various aspects of airport operations (e.g. landside operations and airside operations). Additionally, there are multiple Stakeholders involved, including airport and airline operators, government agencies and retailers. Does this affect the safety of the airport system?
- **Railway sector**: Similarly to the aeroplanes and airports, our trains and railway stations are complex systems! The railway system is a complex system, composed of many interconnected subsystems, that will be examined next in more detail. Does this affect the safety of the railway system?

Let us examine the railway system of the past as presented in Fig. 1.4: the railway system was clear and simple and could easily be understood and managed. Nowadays, the railway system is complex: it is a system of complex systems, as presented in Sect. 1.5.

Today we have **systems of systems, and this affects the safety of the railway system**.

Our systems are not alone—there are many interfaces, as will be discussed in Sect. 1.5.

1.4 Getting Familiar with the Basic Terms

1.4.1 Fault–Failure–Hazard–Risk–Safety

"*Fault–Failure–Hazard–Risk*" are terms used in risk management, but people are confused about their differences. Therefore, we begin with the definitions and explanations of those terms.

1. **Fault** is an *abnormal condition or defect at the component, equipment, or subsystem level, which may lead to failure.*
2. **Failure** is the *lack of ability of a component, equipment, subsystem, or system to perform its intended function as designed.* Failure may be the result of one or many faults.
3. **Hazard** is *anything that can cause harm to people, assets or the environment. A condition that could lead to an accident.*
4. **Risk** Risk is a *probability or threat of damage, injury, liability, loss, or any other negative occurrence that is caused by external or internal vulnerabilities, and that may be avoided through preventive action.*
   The term "risk" means the chance that **something "bad" may happen**. **Risk describes a situation in which there is a chance (probability) of loss or danger**.
   The engineering definition of risk is defined as the product of the probability of an event occurring, that is viewed as undesirable, and an assessment of the expected harm from the event occurring.

Risk = Probability of an accident × Consequence in lost money/deaths

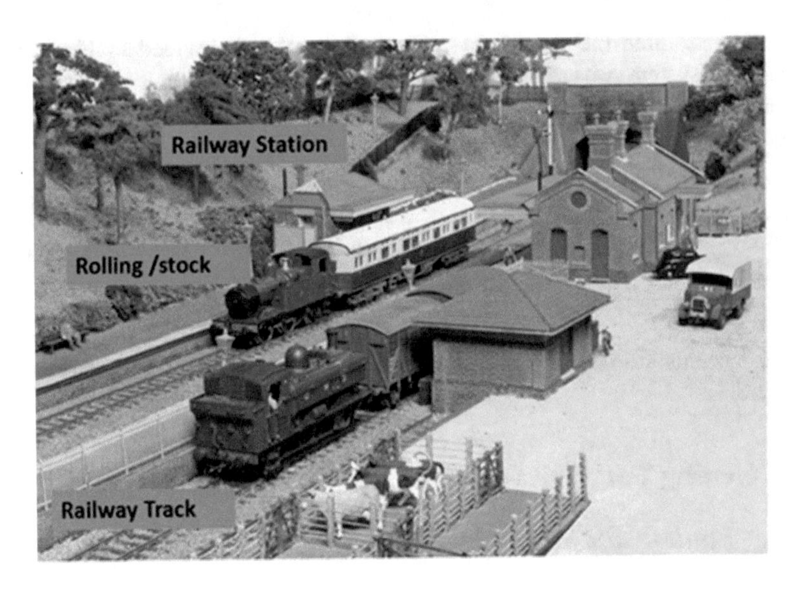

Fig. 1.4 The railway system of the past: a simple system

Risk is related to the considered hazard.
Risk can be thought of as answers to 3 questions:

- What can happen? (scenario)
- How likely is it? (probability/frequency)
- How bad is it? (consequence)

5. **Safety** is the freedom from risk, which is not tolerable. The term "safe" or "safety" is often understood as the state of being protected from all hazards. However, this is not true: "safe" is rather the state of being protected from recognized hazards that are likely to cause harm. Some level of risk is inherent in products or systems [2].
   The relation of risk to safety will be discussed in Sect. 1.4.3.

Next, we will discuss and try to explain above mentioned terms on two examples:

Example 1: Hazard/Risks Crossing a Road
Let us focus on Fig. 1.5: What are the hazards and risks for crossing the road in a city with many cars running and what are the hazards and risks crossing a country road, where is almost no traffic?

The hazard in both cases is the road because there are conditions that could lead to an accident (running cars). The risk is more significant in the cities than in the countryside because the probability of injury or even death is higher in cities.

Fig. 1.5 Crossing a road in a city and a country road

Example 2: Train Running on a Track with Many Faults
Let us focus now on Fig. 1.6. We have a track of very low quality, with many faults.

What are the hazards and risks?

As we defined HAZARD is anything that can cause harm. RISK is how high the chance that the hazard will harm someone.

The bad quality track with many faults is the hazard. Also, a fast running train presents a risk (running faster than the speed limit). RISK: How significant is the chance that the train derails? The track could be very safe for a speed of 20 km/h and very unsafe for 60 km/h, so if the train is running with a speed of 20 km/h the hazard to derail is still there, but the risk this to happen is very low, while in case of a train running with a speed of 60 km/h the risk of a derailment is high.

Let us continue discussing the **difference of the terms *fault–failure–hazard–risk***, based on a case of a broken rail as Fig. 1.7.

- *Fault*: fault is a broken element (due to incorrect design, faulty material, incorrect production…).
- *Failure*: failure is the lack of the ability of the rail to perform its intended function as designed.

Fig. 1.6 A train running on a low-quality track

Fig. 1.7 Broken rail

- *Hazard*: the broken rail presents a hazard.
- *Risk*: risk if the train runs on a broken rail, an accident may occur.

1.4.2 Typical Hazard Sources

Typical hazard sources and associated hazards we need to consider are shown in Fig. 1.8.

1.4.3 The Relation of Risk to Safety

EN 50126-1 defines safety as the "*freedom from unacceptable risk of harm*", considering all the interactions between a system and its environment. This definition addresses safety in all aspects, incorporating functional and technical safety, health and safety issues and impact of human factors.

Safety is the preservation of life, property and the environment. By taking preventive actions, we can prevent accidents, pollution and destruction.

So, **safety is the state of being safe, free from danger, risk or injury: Safety is the absence of danger**.

Due to the variety of unsafe situations or accidents, safety is not easy to measure. Risk analysis is most often used to calculate the safety component, identifying the risks of a specific situation, determining the occurrence and impact, and calculating the total risk (a deeper analysis follows).

What is the relation between Safety and Risk? As mentioned, **safety is freedom from risk**. In order to increase safety, we must first consider the risks. There is an inverse relationship between safety and risk. When one goes up, the other goes down. To increase safety we need to lower the risks (Fig. 1.9).

Because of safety is directly related to risk, **we can manage the safety of our systems by managing the risks.**

... remember that this is not enough!

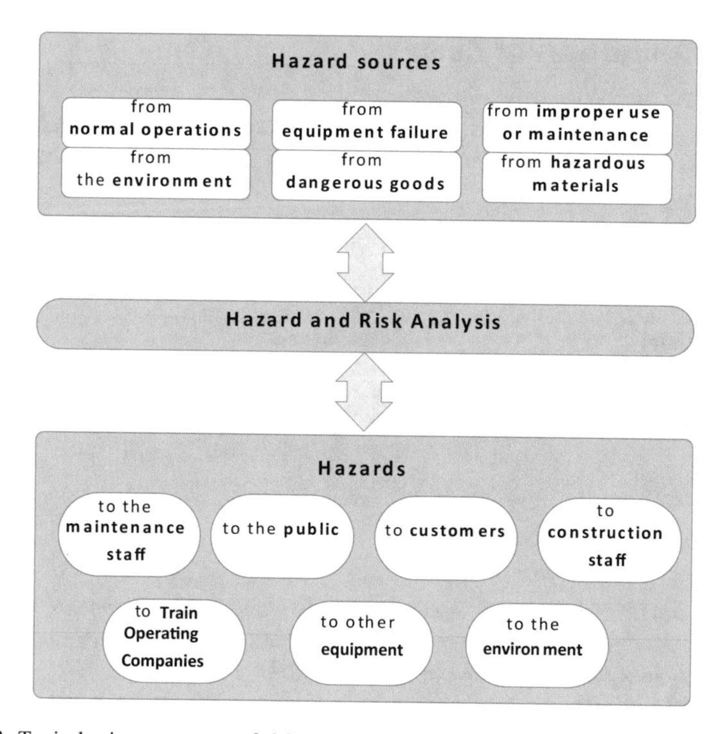

Fig. 1.8 Typical primary sources of risks and hazards

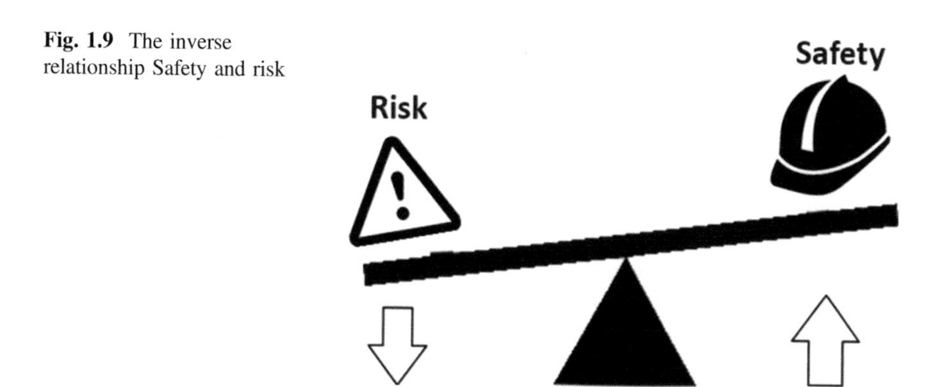

Fig. 1.9 The inverse relationship Safety and risk

To manage the safety a systematic approach is necessary!

The risk management process shall start from a definition of the system under assessment, as will be discussed in Sect. 1.5.

1.4.4 Comparisons of Terms

Following table provides a comparison of terms used in this book with terms used worldwide. Terms mentioned to the left and the right are identical or have a very similar meaning.

• Likelihood	• Probability
• Consequence	• Impact
• Hazard	• Threat
• Risk treatment	• Controlling risk • Risk reduction (ISO Guide 51) • Risk response (PMI) [3, 4] • Risk mitigation (PMI)

1.5 The Railway System

A system can be defined as an "*assembly of subsystems and components, connected in an organised way, to achieve specified functionality*".[3] Certain functionalities are assigned to subsystems and the components within a system.

If the functionality of a subsystem or component changes, the behaviour and state of the system is also changing. Every system is receiving inputs and responds to them producing specified outputs, as per the design. All this happens while our systems are interacting/interfacing with their environment and neighbouring systems.

The interfaces of the railway system must be all the time considered, as they are related to the functionality and the safety of the railway system.

When a system interacts with other systems there is an interface. Figure 1.10 shows some sub-systems: rolling stock, track, signalling, and platform of the railway station, as also the interfaces: an interface occurs whenever a system interacts with other systems i.e. the wheels interact with the tracks and the train doors interact with the passengers. The interfaces are marked with an oval shape. Another interface is shown, between the train driver and the signalling system (Fig. 1.11). The train driver has to follow the indications of the signal: he must stop the train if the signal is red.

These are very simple examples of interfaces, in complex systems, there are hundreds or even thousands of interfaces.

Next Fig. 1.12 presents a broader picture of the railway system. On the tracks, equipped with signalling system, trains are running, which are maintained in depots (equipped with the necessary equipment). Proper traffic management systems regulate traffic (train operation). Trains are stopping at stations, equipped with the

[3]EN 50126-1 (2017) defines system as "set of interrelated elements considered in a defined context as a whole and separated from their environment".

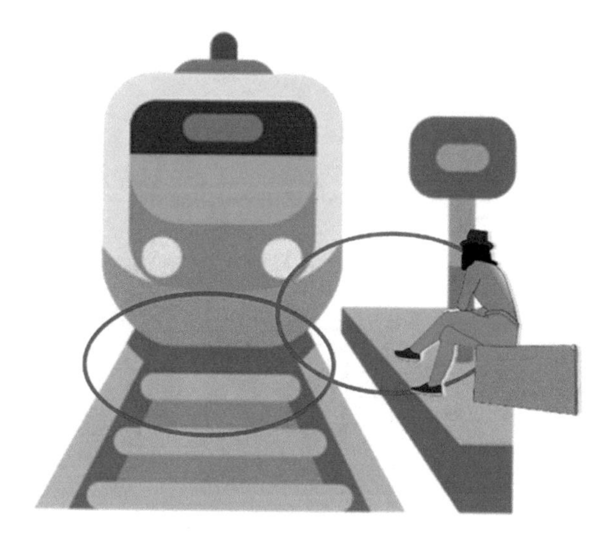

Fig. 1.10 Interfaces in the railway system (1)

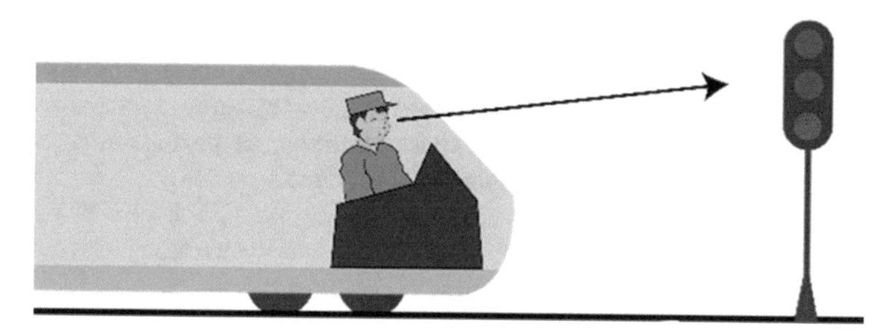

Fig. 1.11 Interfaces in the railway system (2)

necessary facilities. Electric trains are powered with electric power provided by power stations. All the above described are independent subsystems which are interfering with each other and are parts of the railway system.

1.6 Establishing a Safe Railway System: The Big Picture

In this section, we will discuss the parameters that are important for establishing a safe railway system—the big picture- and show how and why risk management is as an important factor for a safe and efficient railway system.

Governments invest in railway systems that have to be efficient; an efficient railway maximises revenues and minimises costs while providing the desired level

Fig. 1.12 Basic parts of the railway system

of service. What are the requirements for an efficient Railway System? It must be **safe**, **secure** and provide **operational excellence**.

As already discussed, our transport systems are complex, but of course, we want to travel safely. In order that to happen, **our complex transport systems must be reliable, available when needed, with short maintenance periods and safe**. So, how can be assured that they will function as required?

To answer this question, let us start by examining the "life cycle" of our subsystems. The "life cycle" of our railway system and subsystems has many phases, but we will consider here four basic phases (Fig. 1.13):

- Subsystem definition and design
- Subsystem construction/manufacturing
- Subsystem maintenance and
- Subsystem operation

Our railway subsystems have to be designed and manufactured as per the specified requirements; the RAMS[4] requirements. Then, the railway subsystems must be approved (placed in service), so we are sure that they meet the system requirements.

After their approval, the sub-systems can be put or placed in service.

How can we assure that the Infrastructure Manager and the Train Operating Company (Railway Undertaking) can manage and operate the subsystem in a safe way? They must elaborate and implement a proper Safety Management System (SMS). Based on their SMS, they can get a Certification. In the European Union, the Infrastructure Manager gets a Safety Authorization. If the Train Operating

[4]RAMS: Reliability, Availability, Maintainability and Safety.

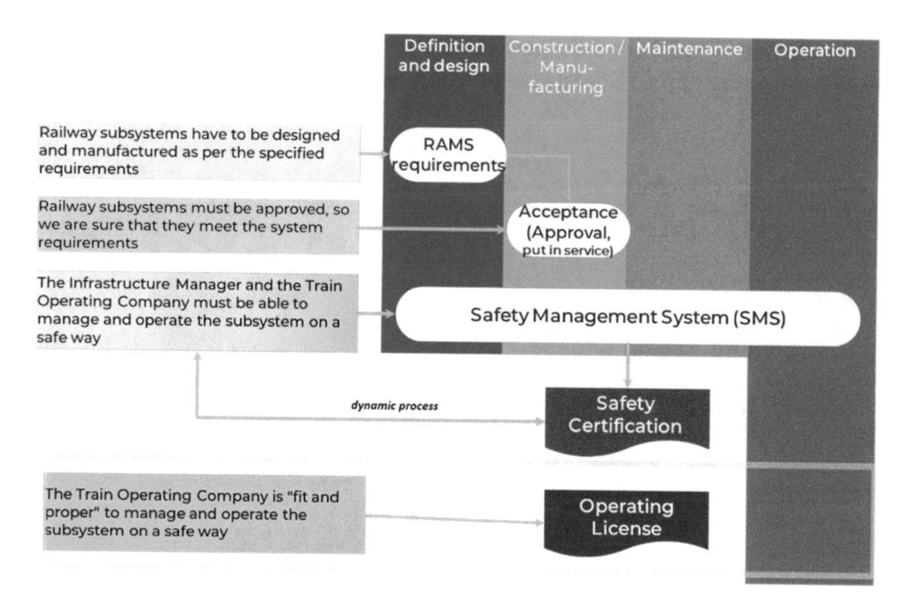

Fig. 1.13 A process to assure that the railway system will function as required

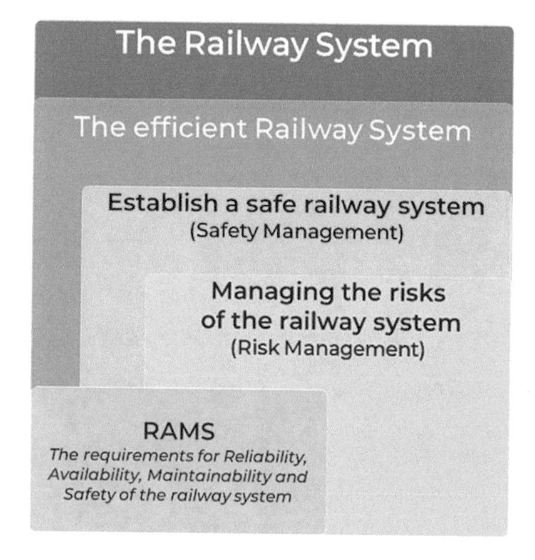

Fig. 1.14 Requirements for the establishment of an efficient Railway System

Company is "fit and proper" to manage and operate the subsystem on a safe way, then, it gets an Operating License.

Let us summarise: to establish an efficient Railway System we need (Fig. 1.14)

- A safe railway system, so we have to manage the safety
- To manage the safety, we have to manage the risks of the railway system
- Finally, the RAMS requirements for Reliability, Availability, Maintainability and Safety of the railway system need to be always considered.

As already mentioned, this book deals with the management of the risks to control the safety of the railway system: **controlling safety is about managing the risks**.

References

1. Reilly J (2017) A short history of risk management. In: Risk management in underground construction conference, Washington DC, 28 Nov 2017
2. ISO/IEC GUIDE 51:2014 Safety aspects—Guidelines for their inclusion in standards
3. CENELEC EN 50126-1:2017—Railway applications. The specification and demonstration of reliability, availability, maintainability and safety (RAMS). Generic RAMS Process
4. PMI (Project Management Institute) (2017) A guide to the project management body of knowledge (PMBOK® Guide), 6th edn

Chapter 2
The Concept of Risk Management

> Life is inherently risky. There is only one big risk you should avoid at all costs, and that is the risk of doing nothing. (Denis Waitley)

2.1 Risk Management

> A decision that does not involve risk, probably is not a decision. (Peter Drucker (a guru of modern management))

2.1.1 General

As we discussed in Sect. 1.1, risks strictly mean "bad news", however, some risks can be positive. Our Organizations are faced with situations (or events) that offer opportunities for benefits (positive risks) or threats to their success (negative risks). Regarding safety, it is generally recognised that consequences are only negative and therefore, the management of safety risk is focused on prevention and mitigation of harm.

Risk management is a key element of all Organizations' strategic management. Risk management is the process by which Organizations approach the risks associated with their activities methodically. Proper risk management focuses on these risks being identified and treated. Proper risk management increases the likelihood of success and helps Organizations achieve their overall goals.

K. Tzanakakis, *Managing Risks in the Railway System*, Springer Tracts on Transportation and Traffic 18, https://doi.org/10.1007/978-3-030-66266-0_2

Organizations are not immune to risks, and their risks must be managed. Risk management should be a process of continuous development. It needs to be integrated into the Organization's culture; it should be part of the Organization's "DNA". The strategy of an Organization must be translated into tactical and operational goals, assigning responsibility across the Organization. Both managers and employees are responsible for managing the risks, and risk management shall be part of their job description.

Risk management refers to the coordinated activities that an Organization takes to direct and control risk.[1]

Risk management can improve the value or protect value, or both. *Value improvement* is the actions, processes and controls put in place to manage the risks affecting the achievement of the Organization's strategy; they increase the potential for strategic outcomes that add value to the Organization.

Value protection is the actions, processes and controls implemented to manage risks which have a negative impact; they protect the Organization's value by preventing or minimizing the impact of negative events.

Risk management helps organizations become more efficient and effective through improved planning and critical thinking and more informed decision-making.

As we discussed, the ultimate objective of risk management is to direct and control the risks of an Organization. There are specified set of processes defined in a risk management framework that support this objective: the Organization's risk management system.

Since incidents such as the Bhopal disaster, and Exxon Valdez incident (see Sect. 1.2), risk management has become an essential management activity.

Organizations should pay attention continuously on their risks. Audits, reviews, and other forms of monitoring are essential for effective risk management.

To be mentioned that low probability risks that don't happen very often are easily forgotten, and their negative characteristics are often left for another day. Consequently, risk management can quickly become ineffective, unless the Organization is taken care about.

Key indicators of an effective risk management activity in an Organization are [1]:

1. The commitment of senior management
2. Risk controls and programs that can be found everywhere in the Organization and are well understood
3. A well-documented "risk profile" that sets risk control priorities
4. Effective risk communication that results in transparency for employees and other Stakeholders, and
5. Monitoring, review, and performance indicators of the Organization's risks. These include all legal and regulatory requirements.

[1]As per ISO 31000, Risk management refers to a "coordinated application of resources to minimise, monitor, and control the probability and/ or impact of unfortunate events or to maximise the realisation of opportunities".

Effective risk management must produce a net value for the Organization. This value is estimated and reviewed and consists of three basic elements:

- costs,
- financial benefits, and
- trust and respect of Stakeholders and the public.

2.1.2 *Do We Need Risk Management?*

As we discussed in previous sections, risk is inherent in everything we do, in our day to day life. Risk is surrounding us. Risk is present in every aspect of our lives, whether we are crossing a road, driving our car, taking the aeroplane or the train, managing a project, purchasing new technological systems, making decisions about the future or deciding not to take any action at all.

Living with risks means that we are managing them continuously, sometimes consciously and sometimes without realizing it. All Organizations and all individuals need to manage risks systematically. This fundamental importance needs to be recognized by all managers and employees.

The alternative to risk management is risky management. Risky management is unlikely to ensure desired outcomes [2].

Risk management is an integral part of good business and quality management practices. Risk management techniques provide a systematic approach for people at all levels to manage risks that are integral parts of their responsibilities.

2.1.3 *Risk Management and Corporate Governance*

Corporate governance is the "*system of rules, practices and processes by which a company is directed and controlled*"[2] (Fig. 2.1). Corporate governance involves essentially balancing the interests of the many Stakeholders of an Organization, such as shareholders, management, customers, suppliers, financiers, government and community. Because corporate governance also provides the framework for achieving the goals of an Organization, it covers almost every area of management, from action plans and internal controls to performance measurement and corporate disclosure.

Risk management promotes good corporate governance by ensuring that top management and stakeholders meet the organizational goals within a tolerable degree of residual risk.

Effective risk management also provides some protection for directors and managers in case of unexpected and unwanted events. In their defence, directors and managers can also prove that they acted with good diligence.

[2]https://www.icsa.org.uk/about-us/policy/what-is-corporate-governance.

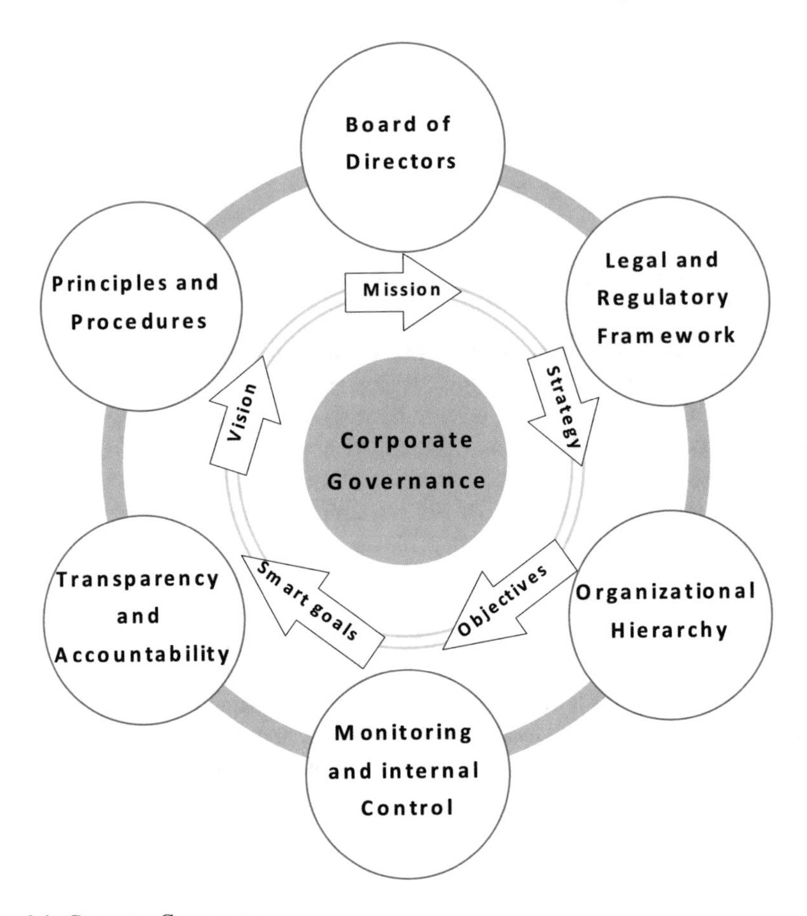

Fig. 2.1 Corporate Governance

Last but not least, risk management provides a structure for facilitating communication and consultation between internal and external Stakeholders (including governing bodies, management and staff at all levels) on the definition and achievement of organizational objectives.

2.1.4 The Risk Management Framework—An Overview of the Different Approaches

A risk management framework effectively integrates the process of risk management within the overall governance, strategy and planning, management, monitoring processes, policies, principles and culture of an Organization.

A risk framework [3]

- is a list of strategic options for the Board to consider
- sets out the risks associated with the risk capacity of the Organization
- sets a risk profile for the Organization when executing its strategy
- provides the Board with additional "horizon scanning" capabilities
- acts as a toolkit to monitor risks.

A good risk management framework should enhance and improve risk management by

- Making it more transparent and clearer to stakeholders
- making the processes more efficient and
- sharing best practices in risk identification, risk assessment and risk treatment implementation.

The framework for risk management needs to be designed to identify measure manage monitor and report the significant risks to achieve the business objectives.

The basic high-level elements of the risk management framework are presented in Fig. 2.2.

In the next sections, the framework for risk management applied in different sectors and different countries in Europe, the American continent and Australia/New Zealand as also at different Organizations will be examined.

2.1.5 The Risk Management Framework in the Civil Aviation Sector

(Basic source: [4]).

2.1.5.1 The Traditional Approach

The traditional approach of safety criticality in the civil aviation sector dictated that a component is safety-critical if a single failure of the technical system in which it is integrated, leads to a catastrophic accident.

The aeroplane technical systems were in principle evaluated with the use of

(a) exclusively the "single fault" criterion, or
(b) exclusively the "fail-safe design" concept.

The "Single Fault" Criterion

The single fault criterion is a requirement, according to which a system designed to carry out a defined safety function must be capable of carrying out its mission in spite of the failure of any single component within the system or in an associated system which supports its operation.

Compliance with this criterion should be considered in order to achieve high safety standards.

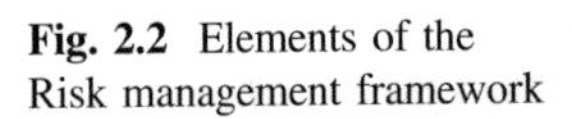

Fig. 2.2 Elements of the Risk management framework

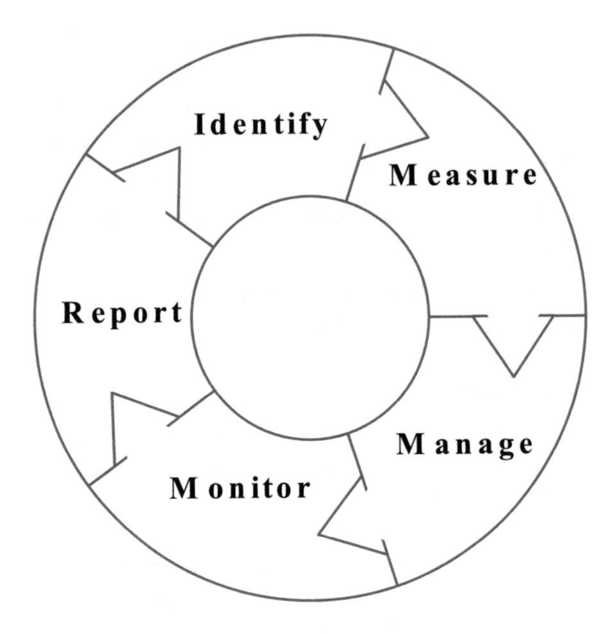

The main disadvantage of this assessment method is that it fails to consider the frequency of failure. Characterizing a technical system as critical to safety would be worthless if there is a low likelihood of a single failure of this system leading directly to a catastrophic accident.

The "Fail-Safe Design" Concept

This concept uses the following design principles or techniques to ensure a safe design:

- Designed integrity and quality, redundancy or backup systems,
- Proven reliability,
- Failure warning or indication,
- Checkability,
- Designed Failure Effect Limits,
- Error-Tolerance, etc.

Regarding this concept, the combination of two or more principles or techniques is usually needed to provide a fail-safe design.

In other words, the system's non-compliance with at least two of them makes it safety-critical.

2.1.5.2 The New Certification Specification "CS-25" Approach[3]

The development of new-generation aeroplanes resulted in a higher level complexity of systems' interaction (highly integrated systems). This led to the need for more complex safety-critical functions. The efficiency of already existing techniques for assessing safety aspects of highly integrated systems was put in question, particularly with the expansion of IT technology and software based techniques.

This led to the development of new approaches, both qualitative and quantitative, for determining safety requirements and establishing compliance with them.

The new approach, known as CS-25, concerning the airworthiness requirements for large aircrafts, covers the design and installation requirements for the aeroplane systems and their associated components, for them to function properly when installed and not to degrade safety.

These requirements are determined by considering both the likelihood and severity of the Failure Condition[4] Effects.

2.1.6 The Risk Management Framework as Per ISO 31000

ISO 31000 is a family of risk management standards codified by the International Organization for Standardisation.

The ISO 31000 family includes[5]:

1. *ISO 31000:2018, Risk management—Guidelines*, provides principles, framework and a process for managing risk. It can be used by any Organization regardless of its size, activity or sector.
2. *ISO Guide 73:2009, Risk management—Vocabulary* complements ISO 31000 by providing a collection of terms and definitions relating to the management of risk.
3. *IEC 31010:2009, Risk management—Risk assessment techniques* focus on risk assessment. Risk assessment helps decision makers understand the risks that could affect the achievement of objectives as well as the adequacy of the controls already in place.

The eleven risk principles proposed by ISO are:

1. Risk management creates value.
2. Risk Management is an integral part of the Organization's processes.
3. Risk management is part of decision-making.

[3]https://www.easa.europa.eu/system/files/dfu/CS-25%20Amendment%2018_0.pdf.

[4]A condition having an effect on the aeroplane and/or its occupants, either direct or consequential, which is caused or contributed to by one or more failures or errors, considering the flight phase and relevant adverse operational or environmental conditions.

[5]https://www.iso.org/iso-31000-risk-management.html.

4. Risk management clearly addresses insecurity.
5. Risk management is systematic, structured and programmed.
6. Risk management is based on the best available information (accessible).
7. Risk management is tailored.
8. Risk management takes into account human and cultural factors.
9. Risk management is transparent and inclusive.
10. Risk management is dynamic, iterative and responsive to change.
11. Risk management facilitates continuous improvement.

The relationship between the principles for managing risk, the framework in which it occurs, and the risk management process described in this International Standard are shown in Fig. 2.3.

When implemented and maintained following this International Standard, the management of risk enables an Organization to, for example [5]:

- increase the likelihood of achieving objectives
- encourage proactive management
- be aware of the need to identify and treat risk throughout the Organization
- improve the identification of opportunities and threats

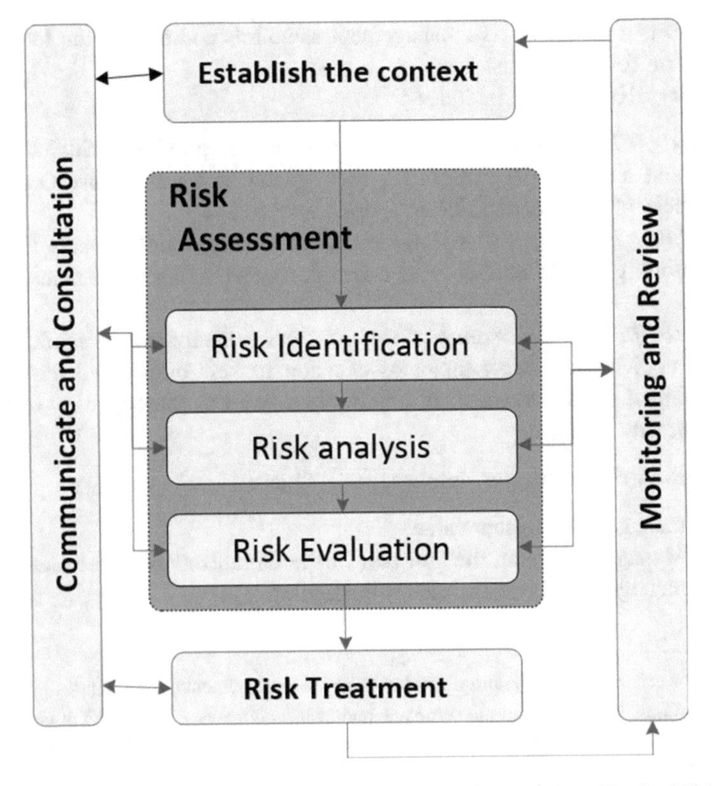

Fig. 2.3 The risk management process as per ISO 31000:2009 (Based on Fig. 1 of ISO 31000)

- comply with relevant legal and regulatory requirements and international norms
- improve mandatory and voluntary reporting
- improve governance
- improve Stakeholder confidence and trust
- establish a reliable basis for decision making and planning
- improve controls
- effectively allocate and use resources for risk treatment
- improve operational effectiveness and efficiency
- enhance health and safety performance, as well as environmental protection
- improve loss prevention and incident management
- minimize losses
- improve organizational learning and
- improve organizational resilience.

2.1.7 The Project Management Institute

The process as provided by the Project Management Institute (see [6, 7]) is focusing on the process of managing projects and is presented next. The steps are as follows:

Plan Risk Management	Defines the scope and objectives of the Project Risk Management process and ensures that the risk process is fully integrated into wider project management
Identify Risks	Identifies as many knowable risks as practicable
Perform Qualitative Risk Analysis	Evaluates key characteristics of individual risks enabling them to be prioritized for further action
Perform Quantitative Risk Analysis	Evaluates the combined effect of risks on the overall project outcome
Plan Risk Responses	Determines appropriate response strategies and actions for each individual risk and overall project risk and integrates them into a consolidated project management plan
Monitor and Control Risks	Implements agreed-upon actions, reviews changes in project risk exposure, identify additional risk management actions as required, and assesses the effectiveness of the Project Risk Management process

Figure 2.4 shows the flow of control and information between the various steps within the Project Risk Management process [6].

2.1.8 The Intergovernmental Organization for International Carriage by Rail (OTIF)

The Intergovernmental Organization for International Carriage by Rail (OTIF) (Fig. 2.5) has adopted risk assessment requirements (UTP GEN-G). The Uniform

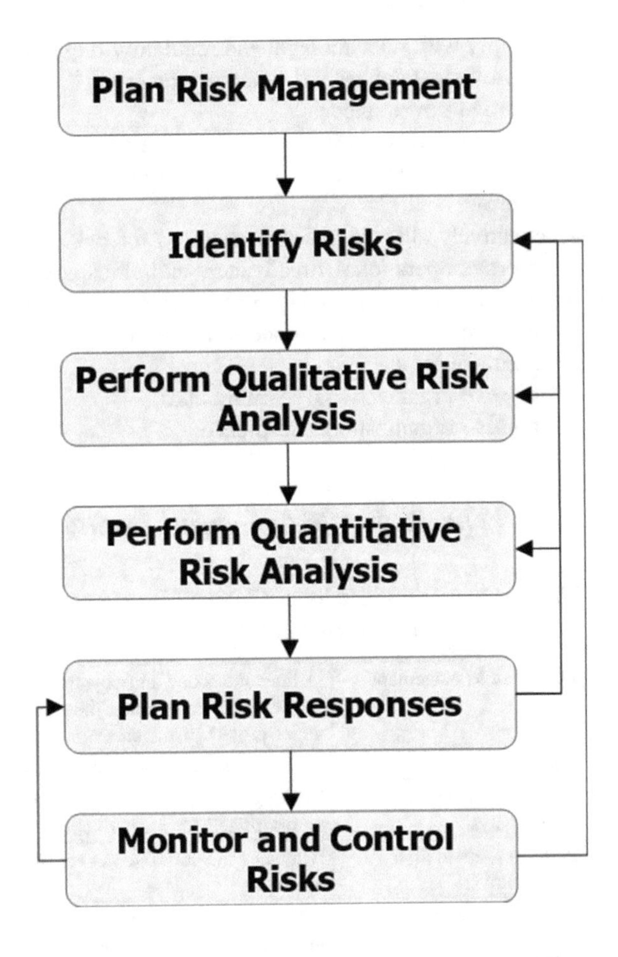

Fig. 2.4 Project risk management process flow diagram (based on [6])

Technical Prescriptions (UTP) are part of COTIF, equivalent to the CSM RA.[6] In principle, each subsystem is subject to one UTP. Where relevant, a subsystem may be covered by several UTP, and one UTP may cover several subsystems.

The UTP GEN-G (version of 1.12.2016), lays down the common safety method (CSM) on risk evaluation and assessment of safety risks of subsystems and the integration of these subsystems into their environment.[7]

---

[6]CSM RA: common safety method (CSM) for risk evaluation and assessment, see Sect. 2.1.12.2.

[7]See https://otif.org/fileadmin/new/3-Reference-Text/3D-Technical-Interoperability/3D1-Prescriptions-and-other-rules/TECH-16037-CTE10-6.4_v2_e_validated_expl-doc-UTP-GEN-G.pdf.

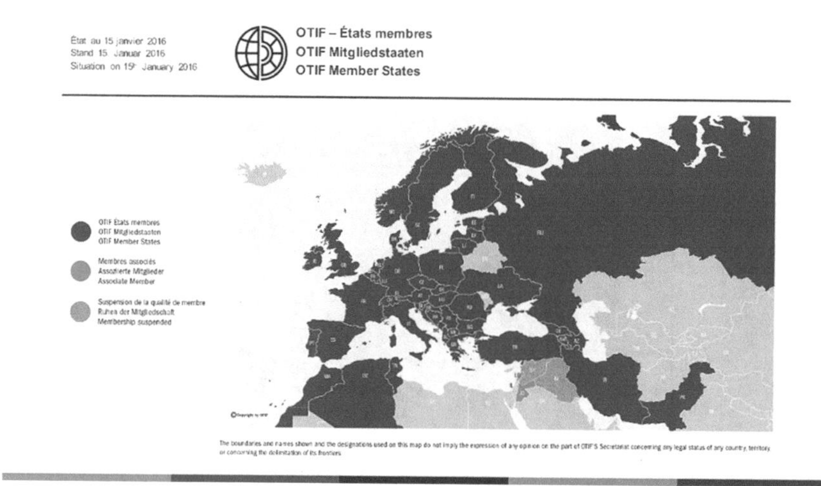

Fig. 2.5 OTIF Members

2.1.9 The Risk Management Framework as Per the Risk Management Standard (UK)

This Risk Management Standard is a **business-based risk management framework**, that is the result of work by a team drawn from the major risk management Organizations in the UK: AIRMIC (The Association of Insurance & Risk Managers), IRM (Institute of Risk management) and The Public Risk Management Association (ALARM).[8] The standard stresses the importance of relating risk management to the Organization's strategic and operational objectives and the threats/opportunities related to achieving those objectives.

As we discussed, the risk is inherent in any decision, at any level in the Organization. As such, the risk management framework closely follows the typical management decision-making structure of:

1. Identify and assess the situation
2. Consider treatment (decision) options
3. Decide
4. Implement management control
5. Monitor decision

Figure 2.6 shows the risk management framework as per the Risk Management Standard [8].

Like all risk management frameworks, it is an **arrangement of processes in a linear sequence**. Feedback is possible at any stage of the process.

[8]The standard is available at http://www.airmic.com/

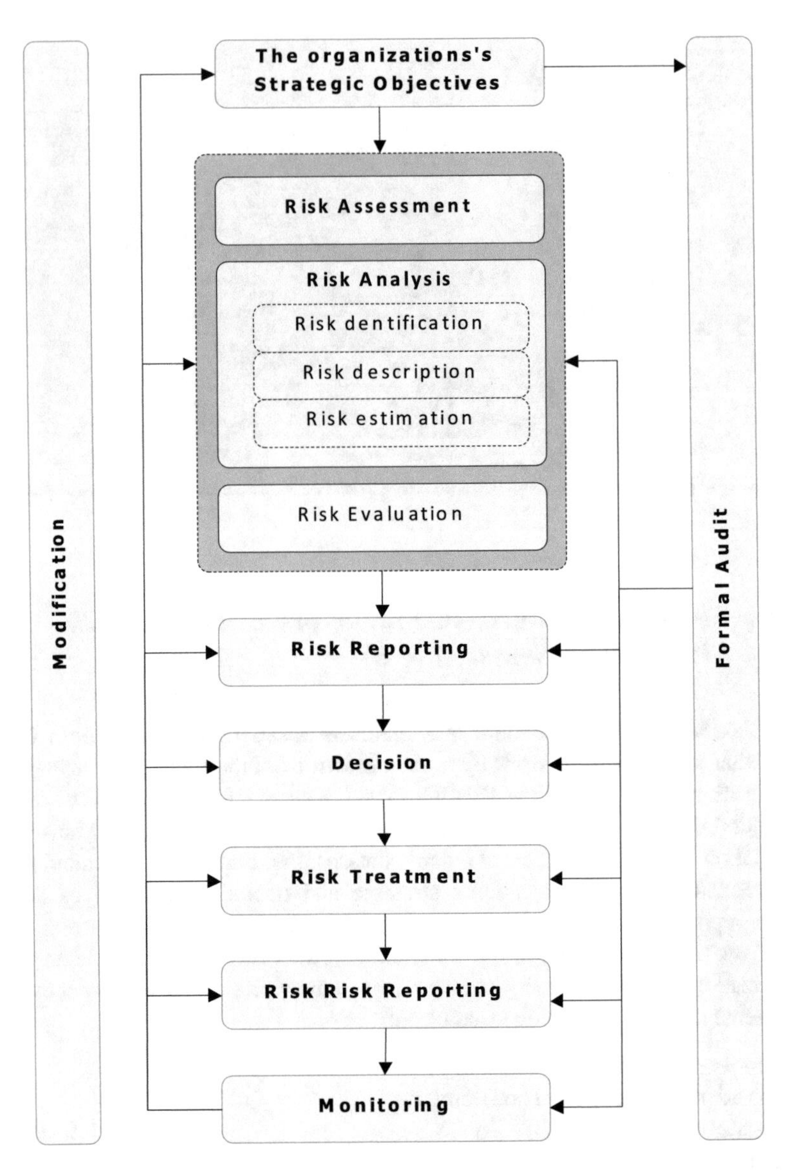

Fig. 2.6 The Risk Management Process as per the Risk Management Standard (UK)

This is a "business" risk management framework that reflects both [1]:

- the traditional health and safety risk management framework and
- the usual management decision processes.

The main stages of the process are:

- the Organization's Strategic Objectives (as expressed in the Risk Management Policy).
- Risk Assessment, composed of Risk Analysis (in turn being Risk Identification, Risk Description, and Risk Estimation) and Risk Evaluation.
- The next step is reporting both opportunities and threats to decision-makers, who must decide if Risk Treatment is needed (i.e. if the risk is acceptable or not with existing Risk Controls),
- then a decision to be taken on the level of Risk Treatment. Decision-making includes discussion of the Residual Risk.
- Finally, there is a formal Audit and Monitoring activity. The framework is unusual in that formal risk communication, and the role of the Stakeholders is not formally recognized.

The standard in Fig. 2.6 uses a risk description method that is similar to the Australian-New Zealand standard described in Sect. 2.1.16.

The "business" framework in Fig. 2.6 differs from the traditional "technical" risk management framework in that it explicitly considers "residual risk reporting" so that risk financing,[9] transfer of risk through insurance, and other residual risk issues can be considered.

2.1.10 *The Enterprise Risk Management (ERM)*

In the last decade, risk management has transformed from the traditional silo mentality[10] to a holistic, co-ordinated and integrated process which manages risk throughout the Organization. This integrated approach has become known as ERM.

ERM is a term used by COSO[11] which published the COSO Enterprise Risk Management—Integrated Framework in 2004 [9]. This has become a well-known framework on how to implement ERM[12] [10, 11].

---

[9]According to ISO definition, in some industries, risk financing refers to funding only the consequences related to the risk and not the risk management.

[10]Silo mentality is an attitude that is found in some organizations; it occurs when several departments or groups within an organization do not want to share information or knowledge with other individuals in the same organization.

[11]The Committee of Sponsoring Organizations of the Treadway Commission (COSO) is a joint initiative of five private sector organizations and is dedicated to providing thought leadership through the development of frameworks and guidance on enterprise risk management, internal control and fraud deterrence.

[12]COSO was not the first to publish practical guidance on an enterprise wide approach to risk management. The first edition of the joint Australian/New Zealand Standard for Risk Management was published in 1995. A further edition, published in 1999, provides guidance on how to establish and implement an enterprise wide risk management process.

The ERM can be defined as the:

> … process effected by an entity's board of directors, management and other personnel, applied in strategy setting and across the enterprise, designed to identify potential events that may affect the entity, and manage risk to be within its risk appetite, to provide reasonable assurance regarding the achievement of entity objectives. (Enterprise Risk Management—Integrated Framework, the Committee of Sponsoring Organizations, COSO, 2004).

To be mentioned that there is no universally agreed definition and the definition above is just one of several definitions developed for Enterprise Risk Management. For example, see the Australian/New Zealand Risk Management Standard 4360 ([12] and Sect. 2.1.16).

The key underlying principles of ERM include:

- consideration in the context of business strategy
- it is everyone's responsibility, with the tone set from the top
- focused strategy, led by the board
- active management of risk
- creation of a risk-aware culture
- a comprehensive and holistic approach to risk management
- consideration of a broad range of risks (strategic, financial, operational and compliance)
- implementation through a risk management framework or system.

As per the COSO—ERM Framework, Enterprise risk management is:

- A process, ongoing and flowing through an entity
- Effected by people at every level of an Organization
- Applied in strategy setting
- Applied across the enterprise, at every level and unit, and includes taking an entity-level portfolio view of risk
- Designed to identify potential events that, if they occur, will affect the entity and manage risk within its risk appetite
- Able to provide reasonable assurance to an entity's management and board of directors
- Geared to the achievement of objectives in one or more separate but overlapping categories

Components of Enterprise Risk Management

Enterprise risk management consists of eight interrelated components. These are derived from the way management runs an enterprise and are integrated with the management process. These components are [11]:

1. **Internal environment** (It includes the risk management philosophy and risk appetite)
2. **Objective setting**
3. **Event identification**

4. **Risk assessment**
5. **Risk response** (to avoid, accept, reduce or share risk).
6. **Control activities** (Policies and procedures help ensure the risk responses are effectively carried out)
7. **Information and communication**
8. **Monitoring** (monitoring the entire ERM process. Modifications made as necessary)

The enterprise risk management framework's structure applies regardless of the size of the institution or how an institution wishes to categorize its risks.

This ERM framework aimed to help managers and Boards of Directors address these relevant business questions:

1. What are all the risks to our business strategy and operations (coverage)?
2. How much risk are we willing to take (risk appetite)?
3. How do we govern risk-taking (culture, governance, and policies)?
4. How do we capture the information we need to manage these risks (risk data and infrastructure)?
5. How do we control the risks (control environment)?
6. How do we know the size of the various risks (measurement and evaluation)?
7. What are we doing about these risks (response)?
8. What possible scenarios could hurt us (stress testing)?
9. How are various risks interrelated (stress testing)?

Table 2.1 provides a comparison of the processes between COSO ERM and ISO 31000.

2.1.11 *The Projects Risk Management Framework in the UK*

The British Standard "*BS 6079-3:2000/ Project management. Guide to the management of business related project risk*" is related to project management. It provides guidance on the identification and control of business-related risks that are coming up when undertaking projects: it is written for project sponsors and project managers.

The standard is similar to other risk management frameworks, particularly the Australian New Zealand standard [12] and the UK standard [13, 14] (Fig. 2.6). It is basically a generic "business" standard. This framework presents clearly **how risk management can, in a general way, be organized in a complex Organization**. The standard also presents the **ALARA** (As Low As Reasonably Achievable) **principle** for positive as well as negative consequences of risks (Fig. 2.7).

Table 2.1 Comparing process activities between COSO ERM and ISO 31000 [22]

COSO ERM	ISO 31000
Internal Environment	Internal and external environment
Objective setting	Identifying and describing objectives
Risk Assessment: • Event identification • Risk assessment • Risk response	Risk Assessment: • Risk identification • Risk analysis • Risk evaluation
Control activities	Risk Treatment
Information and communication	Communication and consultation
Monitoring activities	Monitoring and review

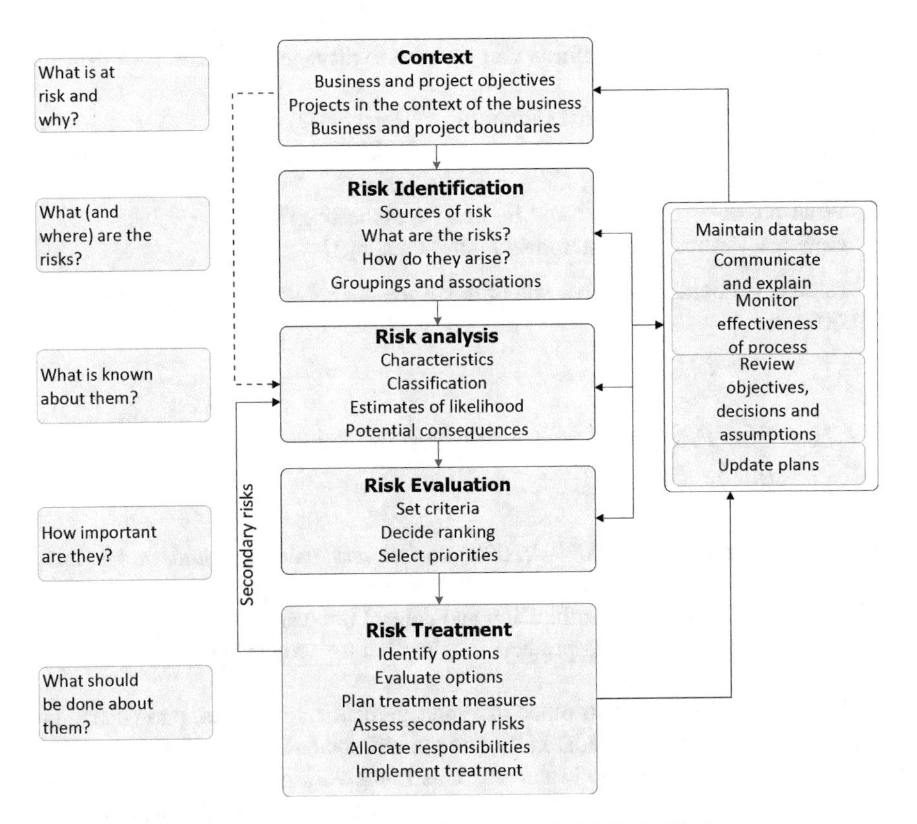

Fig. 2.7 The risk management process (British Standard "BS 6079-3:2000")

2.1.12 *The Risk Management Framework in the European Union*

2.1.12.1 The Directives Related to Safety

The European Commission introduced in 2004 the European Railway Safety Directive 2004/49/EC[13] in order to develop a common legal, structural and Organizational framework within each European Member State to maintain, promote and improve the safety in the rail sector.

The **Railway Safety Directive 2004/49/EC** gives the global legal framework of rail safety in the European Union. Its purpose is to ensure the development and improvement of safety on the European Community's railways and improved access to the market for rail transport services.

This legal framework also defined several common principles:

- The **Common Safety Targets (CST)** which are the acceptable safety levels that must at least be reached by different parts of the rail system. The CSTs shall establish the minimum safety levels to be reached by the system as a whole, and where feasible, by different parts of the rail system in each Member State and in the Union. The CSTs may be expressed in terms of risk acceptance criteria or target safety levels and shall take into consideration, in particular:
  (a) individual risks relating to passengers, staff including employees or contractors, level crossing users and others, and, without prejudice to existing national and international liability rules, individual risks relating to trespassers
  (b) societal risks.
- The **Common Safety Methods (CSM)** describe how the safety levels, the achievement of safety targets and compliance with other safety requirements are assessed, including, where appropriate, through an independent assessment body, by elaborating and defining inter alias the risk evaluation and assessment methods.[14]

[13] https://eur-lex.europa.eu/LexUriServ/LexUriServ.do?uri=OJ:L:2004:220:0016:0039:EN:PDF.

[14] Next, an overview of the regulations related to the common safety method (CSM) for risk evaluation and assessment (CSM RA) is given:

• Commission Implementing Regulation (EU) 402/2013 (the Regulation on a common safety method (CSM) for risk evaluation and assessment (or 'the CSM RA')) came into force on 30 April 2013. It is a framework that describes a common mandatory European risk management process for the rail industry and does not prescribe specific tools or techniques to be used.

• Commission Regulation (EC) 352/2009, which was in force since 1 July 2010, was repealed on 21 May 2015 when Implementing Regulation (EU) 402/2013 started to apply. However, the provisions of 352/2009 will continue to apply in relation to projects which are at an advanced stage of development.

• Commission Implementing Regulation (EU) 2015/1136 was adopted by the European Commission on 13 July 2015 and amends Implementing Regulation (EU) No 402/2013.

- The **Common Safety Indicators (CSI)** which are the indicators enabling to measure the safety performance and the effectiveness of Safety Management System, therefore, they are key- characters of the railway system;
- The **Safety Management System (SMS)**, which is an Organization's formal arrangement, through the provision of policies, resources and processes, to ensure the safety of its work activity. Each Infrastructure Manager and Railway Undertaking (Train Operating Company) in Europe shall implement an SMS.

The **Directive (EU) 2016/798** is a **recast of the Railway Safety Directive** and sets out:

- The risk assessment process to be applied in case of technical, operational or organizational changes
- The criteria to be fulfilled by the assessment body responsible for checking the correct application of the risk assessment process and the results of this application, and the requirements for the accreditation or recognition of its competence to achieve a similar quality of independent assessment
- The harmonised design targets for technical systems which should help with mutual recognition of those systems across the European Union.

For more information, following links are provided:

1. Commission Regulation (EC) 352/2009: http://eur-lex.europa.eu/LexUriServ/LexUriServ.do?uri=OJ:L:2009:108:0004:0019:EN:PDF
2. Commission Regulation (EU) 402/2013 (consolidated with Regulation (EU) 2015/1136): http://eur-lex.europa.eu/legal-content/EN/TXT/PDF/?uri=CELEX:02013R0402-20150803&qid=1486726327105&from=EN
3. Commission Implementing Regulation (EU) 2015/1136: http://www.era.europa.eu/Document-Register/Pages/Commission-implementing-R.aspx
4. The EU Agency for Railways—guidance to the application of the CSM: http://www.era.europa.eu/Document-Register/Documents/guide-for-application-of-CSM-Ver-1-1.pdf
5. The EU Agency for Railways—collection of examples of risk assessments and some possible tools: http://www.era.europa.eu/Document-Register/Documents/collection_of_RA_Ex_and_some_tools_for_CSM_V1.1.pdf
6. The EU Agency for Railways—guidance on harmonised design targets: http://www.era.europa.eu/Document-Register/Pages/Commission-implementing-R.aspx
7. RSSB (Rail Safety Standards Board, UK) guidance: Taking Safe Decisions: http://www.rssb.co.uk/risk-analysis-and-safety-reporting/risk-analysis/taking-safe-decisions

2.1.12.2 Common Safety Method for Risk Evaluation and Assessment (CSM (RA))

The CSM RA[15] applies when any **technical, operational or Organizational change** is being proposed to the railway system. A person making the change ("the proposer") needs first to consider if a change has an impact on safety. If there is no impact on safety, the risk management process in the CSM RA need not be applied, and the proposer must keep a record of how it arrived at its decision.

If the change has an impact on safety, the proposer must decide on whether it is significant or not by using criteria in the CSM RA. If the change is significant, the proposer must apply the risk management process. If the change is not significant, the proposer must keep a record of how it arrived at its decision.

This process is summarised in Fig. 2.8.

In addition to structural, operational or Organizational changes, application of the CSM RA may be required

- by a Technical Specification for Interoperability (TSI) when structural sub-systems are constructed or manufactured, or upgraded or renewed; or
- when placing in service a structural sub-system to ensure that it is safely integrated into the existing system.

Structural sub-systems (as described in Directive 2008/57/EC) are:

- rolling stock
- infrastructure
- command control and signalling and
- energy.

2.1.12.3 Applying the Risk Management Process of the CSM RA[16]

The CSM RA applies to "*any change of the railway system in a Member State ... which is considered to be significant within the meaning of Article 4 of the Regulation*". Those changes may be technical, operational or Organizational, but are those who could impact the operating conditions of the railway system.

The proposer[17] of a change is responsible for applying the risk management process set out in the CSM RA.

Figure 2.9 shows the risk management process defined in the CSM RA. The process primarily consists of the following steps:

a. The proposer of a change produces a preliminary definition of that change and the system to which it relates. It then examines it against the significance criteria

[15]Common safety method for risk evaluation and assessment (CSM RA).

[16]CSM for risk evaluation and assessment.

[17]Proposer is a person making the change (technical, operational or Organizational change).

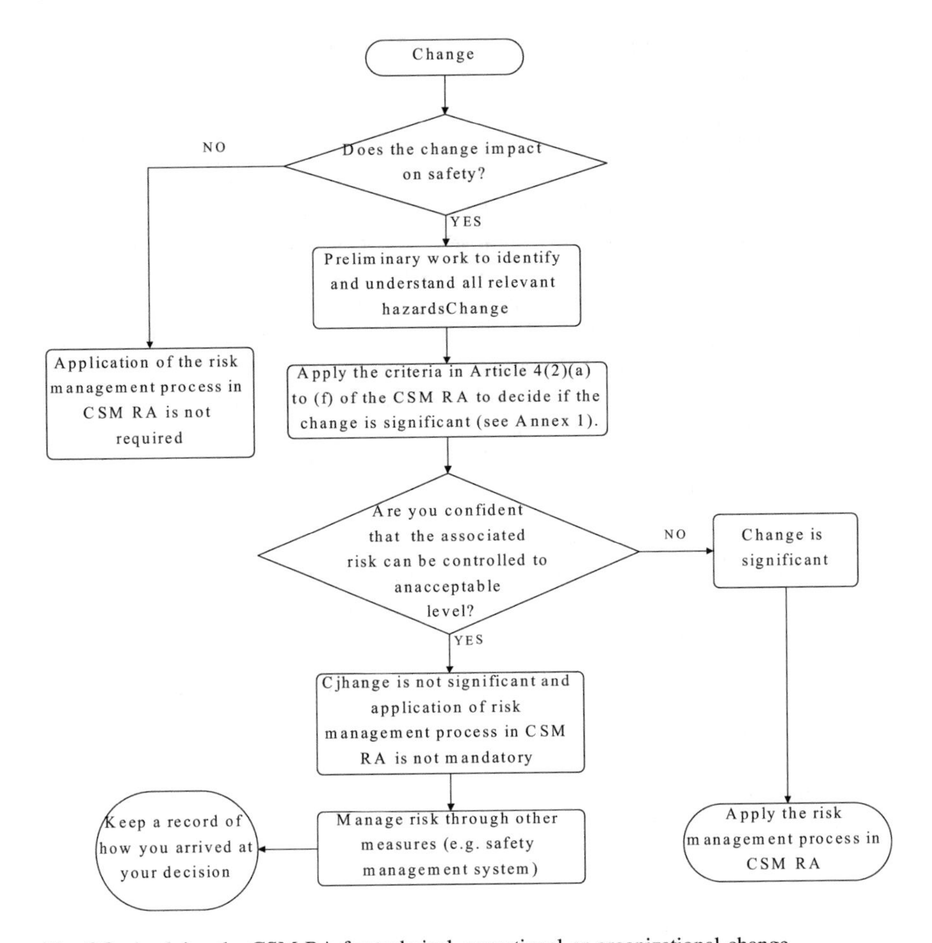

Fig. 2.8 Applying the CSM RA for technical, operational or organizational change

in the regulation. If a change is deemed to be significant, then the regulation requires the proposer to apply the risk management process and appoint an independent assessment body to assess the application of the process. However, the CSM RA risk management process is a sound one and the proposer may choose to apply some or all of it more generally.

b. The CSM, risk management process, starts with the system definition. This provides the key details of the system that is being changed—its purpose, functions, interfaces and the existing safety measures that apply to it. This system definition will be kept alive for the duration of the project.
c. All reasonably foreseeable hazards are identified, and their risk is classified and/or analysed.
d. Safety requirements are identified by application of one or more of the three risk acceptance principles to each hazard.

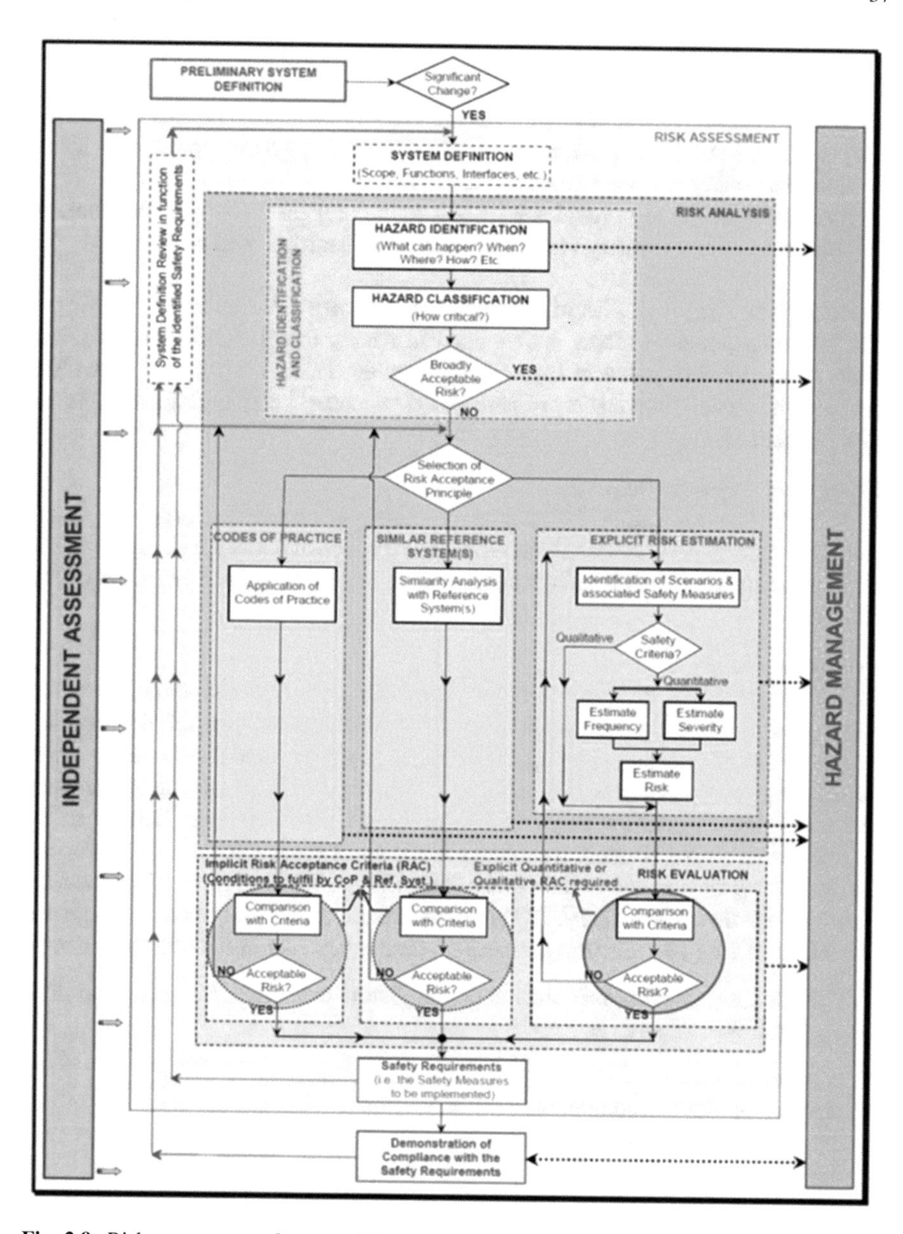

Fig. 2.9 Risk management framework in the CSM Regulation

e. A hazard record for the system that is to be changed is produced and maintained. Its purpose is to track the progress of the project's risk management process.
f. Before acceptance, the change proposer demonstrates that the risk assessment principles have been correctly applied and that the system complies with all specified safety requirements.
g. The assessment body provides its report to the proposer. The proposer remains responsible for the safety and decides to implement the proposed change.

What Are the Main Phases of the Risk Management Process?
The risk management process is contained in Annex I of the CSM RA [15]. The main phases are illustrated in Fig. 2.9, and further details are set out below. The process illustrated is not static or linear as the proposer[18] may undertake iterations of all or part of the process.

Preliminary System Definition
In order to assess whether the change is significant or not, the proposer should conduct a preliminary system definition. This 'preliminary system definition' is in effect an analysis of what is being changed and a preliminary risk assessment of that change.

System Definition
The risk managBasic Sourceement process starts with the system definition (which can use information from the preliminary system definition). This provides the key details of the system that is being changed—its purpose, functions, interfaces and the existing safety measures that apply to it. In most cases, the hazards which need to be analysed will exist at the boundary of the system with its environment.

The definition is not static, and during iterations of the risk management process, it should be reviewed and updated with the additional safety requirements that are identified by the risk analysis. It, therefore, describes the condition (or expected condition) of the system before the change, during the change and after the change.

The system definition may change due to factors other than the specification of safety requirements, such as

- changes in scope
- changes in client requirements
- increasing design definition and
- implementation of changes proposed by contractors and suppliers.

Such changes may necessitate iteration of the risk management process.

Equally, changes to the system definition for other reasons may require the proposer to repeat all or part of the process and discuss with the assessment body the implications.

[18]A proposer is a person making the change.

The risk management process states that the system definition should address at least the following issues:

a. system objective i.e. intended purpose
b. system functions and elements, where relevant (including, i.e. human, technical and operational elements)
c. system boundary including other interacting systems
d. physical (i.e. interacting systems) and functional (i.e. functional input and output) interfaces
e. system environment (i.e. energy and thermal flow, shocks, vibrations, electromagnetic interference, operational use)
f. existing safety measures and, after iterations, the definition of the safety requirements identified by the risk assessment process and
g. assumptions which shall determine the limits for the risk assessment.

The system definition needs to cover not only normal mode operations but also degraded or emergency mode.

Consideration of interfaces should not be restricted to physical parameters, such as interfaces between wheel and rail. It should include human interfaces, such as the user-machine interface between the driver and driver displays in the cabs of rail vehicles. It should also include interfaces with non-railway installations and Organizations. For example, the interface with road users at level crossings.

Operational rules and procedures and staff competence should be considered as part of the system environment. This is in addition to the more common issues such as weather, electromagnetic interference, local conditions such as lighting levels, etc.

A good test of whether the system definition is complete and sufficient is if the proposer can describe the system elements, boundaries and interfaces, as well as what the system does.

The description can effectively serve as a model of the system and should cover

- structural issues (how the system is constructed or made up); and
- operational issues (what it does, and how it usually behaves and in failure modes).

The existing safety measures, which may change as the risk assessment process progresses, can be added after the structural and operational parts of the model are complete.

For some projects, the proposer may not know all the environmental or operational conditions in which the altered or new system will operate. In these circumstances, they should make assumptions based on the intended or most likely environment. These assumptions will determine the initial limits of use of the system and should be recorded. When the system is put into use, the proposer (who may be different to the original proposer) should review the assumptions and analyse any differences with the intended environmental and operational conditions.

Hazard Identification
The purpose of the hazard identification is to identify all reasonably foreseeable hazards, which are then analysed further in the next steps.

The hazard identification should be systematic and structured, which means taking into account factors such as

- the boundary of the system and its interactions with the environment
- the system's modes of operation (i.e. normal/degraded/emergency)
- the system life cycle including maintenance
- the circumstances of operation (i.e. freight-only line, tunnel, bridge, etc.)
- human factors
- environmental conditions and
- relevant and foreseeable system failure modes.

While the risk management process does not require that any specific tools should be applied, many of the more well-known techniques (see Chap. 4) will be relevant, including

- structured group discussions
- checklists
- task analysis
- hazard and operability studies (HAZOPs)
- hazard identification studies (HAZIDs)
- failure mode and effects analysis (FMEA)
- fault trees and
- event trees.

Whichever technique is used, it is important to have the right mixture of experience and competence while maintaining impartiality and objectivity. Correct hazard identification will underpin the whole risk management process and give assurance that the risks will be managed in the project.

The risk management process uses the term "broadly acceptable" to identify those hazards which need not be analysed further. In this context, 'broadly acceptable' applies to those hazards where the risk is, to all intents and purposes, insignificant or negligible. This could be because the hazard is so unlikely to arise that there are no feasible control measures that could be used to control the risk it creates (i.e. earthquakes if in a low vulnerability area) or where there is a credible failure mode, but the consequences are negligible. By screening out the 'broadly acceptable' hazards at this stage, the risk analysis can focus on the more important hazards to manage. It is unlikely that many hazards will be screened out in this way.

The level of detail of the hazard identification depends on the system that is being assessed and needs to be sufficient to ensure that relevant safety measures can be identified. If following a high-level hazard identification, it can be successfully demonstrated that a hazard can be controlled by the application of one of the three risk acceptance principles required by the risk management process, then no further hazard identification is necessary. If it is not possible to have sufficient confidence at this stage, then further analysis of the causes of these high-level hazards is undertaken to identify relevant measures to control the risks arising. The risk management process continues until it can be shown that the overall system risk is controlled by one or more of the risk acceptance principles.

Hazard identification is still necessary for those changes where the hazards are to be controlled by the application of codes of practice or by comparison to reference systems. Hazard identification in these cases will serve to check that all the identified hazards are being controlled by relevant codes of practice or by adopting the safety measures for an appropriate in-use system. This will also support mutual recognition and transparency. The hazard identification can then be limited to verification of the relevance of the codes of practice or reference systems if these completely control the hazards and identification of any deviations from them. If there are no deviations, the hazard identification may be considered complete.

The purpose of risk analyses and evaluation is to identify those safety requirements and measures that are necessary to control the risks arising from the identified hazards.

Risk acceptance Principles

Hazards can be analysed and evaluated using one or more of the following risk acceptance principles:

- the application of codes of practice
- a comparison with similar systems (reference systems) or
- an explicit risk estimation.

Codes of Practice

Standards and rules have to meet all the following criteria to be used as a code of practice for the risk management process:

- be widely accepted in the railway sector or otherwise justified to the assessment body
- be relevant for the control of the specific hazard and
- be available to an assessment body so that it can:
- assess the suitability of the how the CSM RA is applied and the results of applying it or
- mutually recognise any safety assessment report on the same system.

Reference Systems

Reference systems can be used to derive the safety requirements for the new or changed system. For an existing system to be used as a reference system, a proposer needs to demonstrate that

- it has been proven in use and has an acceptable safety level
- it is accepted in the Member State where the change is to be introduced and
- the system being assessed is used under similar functional, operational and environmental conditions and has similar interfaces as the reference system.

Explicit Risk Estimation

Explicit risk estimation is an assessment of the risks associated with hazard(s), where risk is defined as a combination of the rate of the occurrence of the hazard or hazardous event causing harm (the frequency) and the degree of severity of the harm (the consequence).

The estimation can be qualitative, semi-quantitative or quantitative. The choice will be determined by factors such as the availability of quantitative data and confidence in such data. Any analysis should be proportionate to the potential risks. Any risk assessment should follow a systematic and structured process.

Hazard Record

The proposer has to create and maintain a hazard record for the system (or part system) that is to be changed. Its purpose is to track the progress of the risk assessment and risk management process for the project. The CSM RA requires that it contains specific information but does not mandate any particular format.

Other Documentation

The CSM RA places some minimum requirements on proposers to document certain information to assist the assessment body. These are

- a description of the Organization and the experts appointed to carry out the risk assessment process
- the results of the different phases of the risk assessment and a list of all the necessary safety requirements to be fulfilled in order to control the risk to an acceptable level
- evidence of compliance with all the necessary safety requirements and
- all assumptions relevant for system integration, operation or maintenance, which were made during system definition, design and risk assessment.

Demonstration of System Compliance

The proposer 'accepts' the change in the system and is responsible for its safe integration and operation on the broader railway system. This means ensuring that the system is designed, validated and accepted against the safety measures identified to control the hazards. Before acceptance, the proposer needs to demonstrate that the risk assessment principles have been correctly applied and that the system complies with all specified requirements. The proposer has overall responsibility for coordinating and managing the demonstration that the safety requirements are met. Other Organizations involved will need to demonstrate that they have met the safety requirements and implemented safety measures at the lower level for the part of the system which they are responsible.

Independent Assessment

An assessment body must carry out the independent assessment. The CSM RA requires an independent assessment of

- how the risk management process is applied; and
- the results from the risk management process.

2.1.12.4 Hazard Classification

Hazard classification is based on an initial assessment of the risk associated with each hazard and is carried out as part of a hazard identification process.

The regulation [15] states that:

> To focus the risk assessment efforts upon the most important risks, the hazards shall be classified according to the estimated risk arising from them. Based on expert judgement, hazards associated with a broadly acceptable risk need not be analysed further but shall be registered in the hazard record. Their classification shall be justified in order to allow independent assessment by an assessment body.

Classification of hazards allows the proposer to focus risk assessment on the most important risks, by omitting those hazards which need no further consideration at an early stage of the project. The basis of the classification of a hazard is **expert judgement**. In practice, this is usually done by a hazard identification workshop via the collective opinion of the attendees.[19]

The regulation [15] states that:

> As a criterion, risks resulting from hazards may be classified as broadly acceptable when the risk is so small that it is not reasonable to implement any additional safety measure. The expert judgement shall take into account that the contribution of all the broadly acceptable risks does not exceed a defined proportion of the overall risk. (Annex I, clause 2.2.3)

Regarding the term "broadly acceptable" as per the ORR Guidance *(Clause 3.22)* [16]

> "broadly acceptable" to identify those hazards which need not be analysed further. In this context, "broadly acceptable" applies to those hazards where the risk is, to all intents and purposes, insignificant or negligible. This could be because the hazard is so unlikely to arise that there are no feasible control measures that could be used to control the risk it creates or where there is a credible failure mode, but the consequences are negligible. By screening out the 'broadly acceptable' hazards at this stage, the risk analysis can focus on the more important hazards to manage. It is unlikely that many hazards will be screened out in this way.

The regulation also states that:

> The expert judgement shall take into account that the contribution of all the broadly acceptable risks does not exceed a defined proportion of the overall risk. (Annex I, clause 2.2.3)

Given that the broadly acceptable risks are by definition very low, and that it would not be expected that many hazards would be discounted in this way, it is likely that the contribution of all broadly acceptable risks would be insignificant compared to the overall risk.

[19] A record of who attended the hazard identification workshop, or otherwise took part in the classification, will help demonstrate that the requirement to apply expert judgement has been met.

To be noted that in related standards, such as EN 50126, classification has a broader use as an initial stage of risk assessment, incorporating the use of a risk matrix.

2.1.12.5 Correspondence Between the CSM Risk Assessment Process and the V-Cycle (V-Model, CENELEC EN 50126-1)

(Basic Source: [17]).

The risk management process covered by the CSM can be represented within a V-Cycle (as per CENELEC EN 50126-1 [18]) that starts with the (preliminary) system definition and that finishes with the system acceptance: see Fig. 2.10. This simplified V-Cycle can be mapped then on the classical V-Cycle in Fig. 7 of the EN 50126-1 standard [18] (Fig. 2.11). In order to show the correspondence of the CSM risk management process in Fig. 2.10, the V-Cycle is recalled in Fig. 2.11:

1 the "preliminary system definition" of the CSM in Fig. 2.9 corresponds to the Phase 1 in CENELEC V-Cycle, i.e. to the definition of the system "concept" (see BOX 1 in Fig. 2.11);
2 the "risk assessment" of the CSM in Fig. 2.9 includes the following phases of the CENELEC V-Cycle (see BOX 2 in Fig. 2.11):
   a. Phase 2 in Fig. 2.11: "system definition and application conditions"
   b. Phase 3 in Fig. 2.11: "risk analysis"
   c. Phase 4 in Fig. 2.11: "system requirements"
   d. Phase 5 in Fig. 2.11: "apportionment of system requirements" down to the different sub-systems and components.

The outputs of the risk assessment process in the CSM are (after iterations—see Fig. 2.9):

1 the "system definition" updated with the "safety requirements" issued from the "risk analysis" and "risk evaluation" activities
2 the "architecture and apportionment of system requirements" down to the different sub-systems and components (Phase 5 in Fig. 2.11)
3 the "hazard record" that registers:
   a. all the identified hazards and the associated safety measures
   b. the resulting safety requirements
   c. the assumptions taken into account for the system that determine the limits and validity of the risk assessment
4 and in general, all the evidence resulting from the application of the CSM

These risk assessment outputs of the CSM correspond to the safety related outputs of Phase 4 in the V-Cycle, i.e. to the system requirement specification in Fig. 2.11.

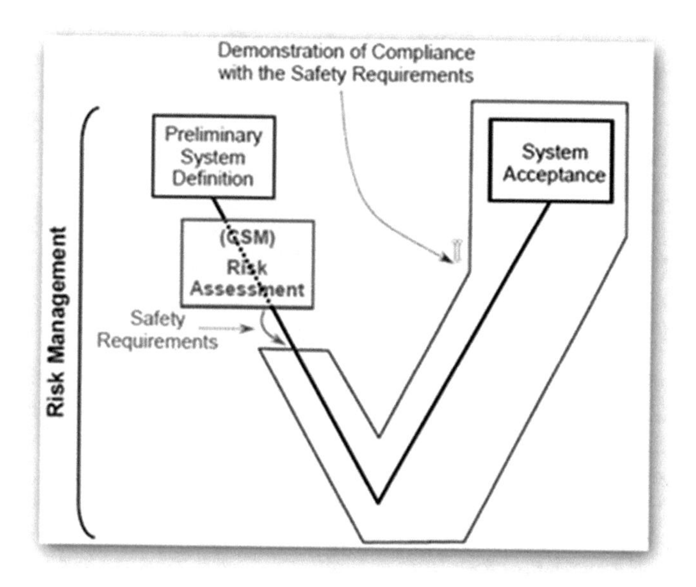

Fig. 2.10 Simplified V-Cycle of Fig. 10 of EN 50126 Standard

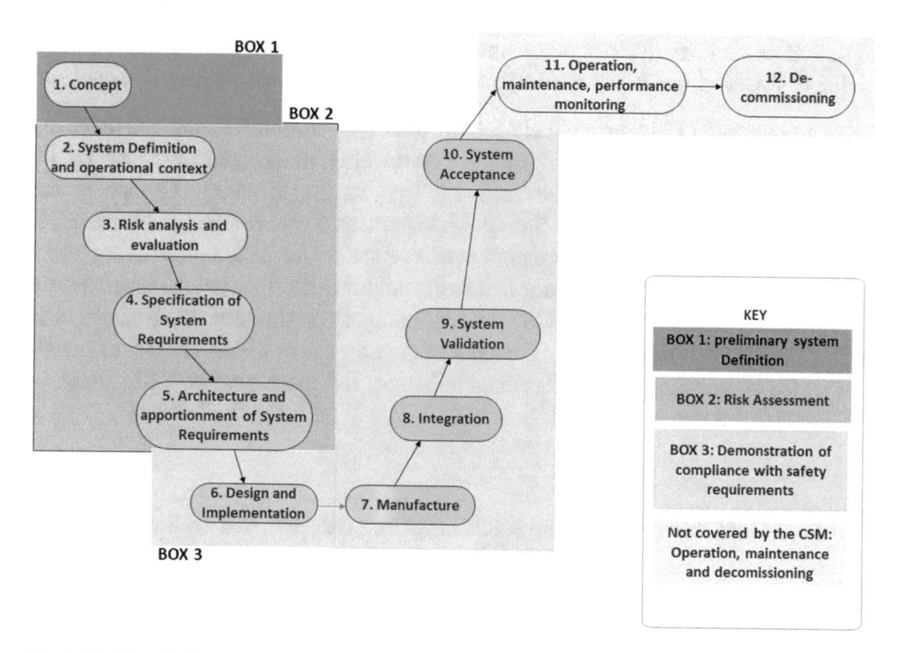

Fig. 2.11 The V-Cycle (as per Fig. 7 of EN 50 126 V-Cycle (system life-cycle) [18])

The system definition updated with the results from the risk assessment and the hazard record constitute the inputs against which the system is designed and accepted. The "demonstration of the system compliance with the safety requirements" in the CSM corresponds to the following phases of the CENELEC V-Cycle (see BOX 3 in Fig. 2.11):

1. Phase 6 in Fig. 2.11: "design and implementation"
2. Phase 7 in Fig. 2.11: "manufacture"
3. Phase 8 in Fig. 2.11: "integration"
4. Phase 9 in Fig. 2.11: "system validation"
5. Phase 10 in Fig. 2.11: "system acceptance".

The demonstration of the system compliance with the safety requirements depends on whether the significant change is technical, operational or Organizational.

The "demonstration of the system compliance with the safety requirements" in the CSM in practice covers all the phases "6 to 10" (see the list here above and Fig. 2.11) in the V-Cycle. They include the design, manufacture, installation, verification and validation activities, as well as the associated RAMS activities and the system acceptance.

It results then from the comparison with the classical V-Cycle in Fig. 2.11 that:

1. the CSM covers the phases "1 to 10" of this V-Cycle. They include the set of activities required for the acceptance of the system under assessment
2. the CSM does not cover the phases "11" and "12" of the system life-cycle:
   a. the phase "11" is respectively related to the "operation and maintenance" and "performance monitoring" of the system after its acceptance based on the CSM. This phase is covered by the Train Operating Company and Infrastructure Managers Safety Management System (SMS). However, if during the operation, the maintenance or the performance monitoring of the system it appears necessary to modify and retrofit the system, whereas it is already in operation, the CSM is applied again on the new required changes.
   b. the "de-commissioning" of a system already in operation (Phase 12) could also be considered as a significant change, and therefore, the CSM could be applied again.

2.1.12.6 Overview of Basic Regulations/Standards Related to Safety

Figure 2.12, provides an overview of basic regulations/standards related to safety that applies in the European Union.

Figure 2.13 presents the development timeline of the basic regulations/standards (Regulation CSM RA and the Euronorms), related to safety.

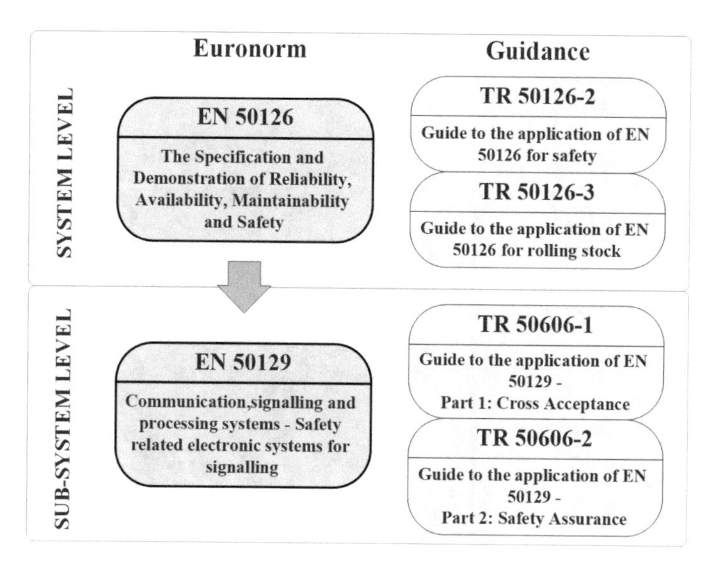

Fig. 2.12 Overview of current railway standards

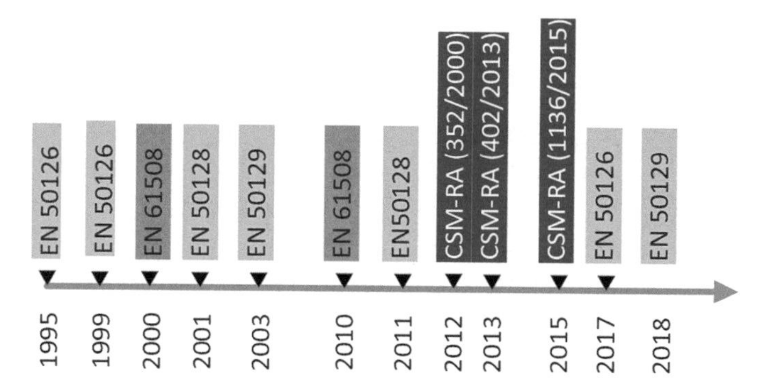

Fig. 2.13 The development timeline for CENELEC/EN and the CSM-RA

2.1.13 The Risk Management Framework in the USA

The US framework is presented in Fig. 2.14. It is circular and has "decision" arrows linking all steps, emphasizing the feedback and unique design aspects of risk management. For every problem and its characteristics, the risk management process needs to be specially designed using standard functional elements.

As per [1], the US approach is characterized as "holistic", concentrating on health outcomes, and emphasizing various risk treatment interventions and strategies that include mixtures of public and private interventions, less dependence on command and control systems, and so on. The US framework has shifted from

Fig. 2.14 The risk management framework in the USA [1]

stressing a prescribed administrative and technical process (as other frameworks) to emphasizing population health outcomes in an open process with stakeholder active involvement.

Stakeholders' involvement is an important element of the US system, they have to be involved in all process steps.

The US framework is in general comparable to the other frameworks for the core aspects of risk assessment, risk treatment, risk communication, decision-making, monitoring, etc. This framework is a more conceptual framework which needs to be operationalized by a specific Organization.

2.1.14 The Risk Management Framework in Japan

Japan issued the standard "*JSI Q 2001:2001—Guidelines for development and implementation of risk management system*", using the ISO terminology for risk management system as well as other terms [1]. Figure 2.15 presents the Japanese framework. It shows the concept of the risk management system as an integral part of an Organization's structure established to maintain risk management activities and the associated system performing those activities.

There are two basic advances in the standard, the first is the formal definition and development of the risk management system (as opposed to the usual development

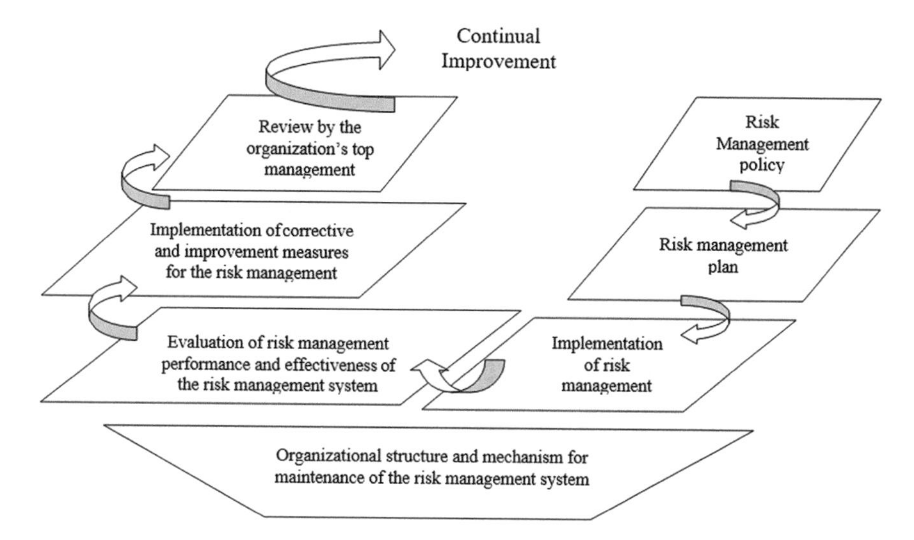

Fig. 2.15 The Japanese Industrial Standard Risk Management System (JSA, 2001) [1]

of the risk management framework) and the second advance is the linkage of the risk management system directly to Total Quality Management concepts, including continuous improvement. In the details of the standard, there is a focus on prevention of the events following the 1995 major earthquake when some 5,000 firms and Organizations were unable to restart operations quickly [1].

2.1.15 The Risk Management Framework in Canada

Treasury Board of Canada issued in 2001 an "*Integrated Risk Management Framework*" to provide generic guidance for risk management frameworks in a variety of government agencies and ministries [1]. The framework consists of 9 steps, or functional activities (translated into ISO terminology):

1. identify risk areas,
2. set the context
3. identify risks
4. estimate risk probabilities and consequences
5. assess risks and select priority risks
6. establish risk criteria
7. risk treatment options; decide on risk treatments (risk evaluation and acceptance)
8. implement risk controls
9. monitor and repeat steps as required.

Figure 2.16 illustrates the Canadian Risk Management Standard (Guide) Q850 (CSA, 1997). It is a generic standard and was the first to formally introduce the ideas of:

1. Explicit decision-making at most steps in the process, including the decision option of "end" consideration of risk.
2. Stakeholders, Stakeholder analysis, and introduction of Stakeholder needs, issues, and concerns in the risk evaluation, etc.
3. Explicit introduction of risk communication at each functional step in the framework.
4. Initiation step where the technical and administrative risk management "process" is designed and resourced. Initiation is a function in the Organization's risk management system.
5. Preliminary Analysis, which is a screening level risk analysis and risk evaluation (both together are called risk assessment).
6. Documentation needs and the creation of a "risk library" for any application of the process.

This standard was a milestone and the Australian-New Zealand standard issued in 1995 was revised in 1999 to incorporate some of these ideas and extend them in terms of the concept of "Context" and other innovations.

The Canadian standard as per Fig. 2.16 now has some non-standard (ISO) terminology, for example, Risk Control is now Risk Treatment after implementation, and Risk Treatment is the process of finding a treatment to modify the risk that the decision-makers deem acceptable.[20] However, most of the terminology is similar to ISO/IEC Guide 73.

2.1.16 The Risk Management Framework in Australia/New Zealand

The Australian/ New Zealand framework [2] separates the assessment of the risk with already existing controls and then if the remaining residual risk is not acceptable, additional treatment options are evaluated to make the residual risk acceptable [1].

Figure 2.17 illustrates the Australian-New Zealand standard framework of functional processes [12]. It has a core set of functions (Fig. 2.17), with slightly different terminology, including the idea of extensive feedback loops and continuous monitoring. Like the Canadian framework, it explicitly identifies the "communicate and consult" function.

---

[20]This is analogous to the field of medicine where treatments are the available therapeutic options for disease management, while risk controls are the specific care plan the physician has selected from the available treatment options.

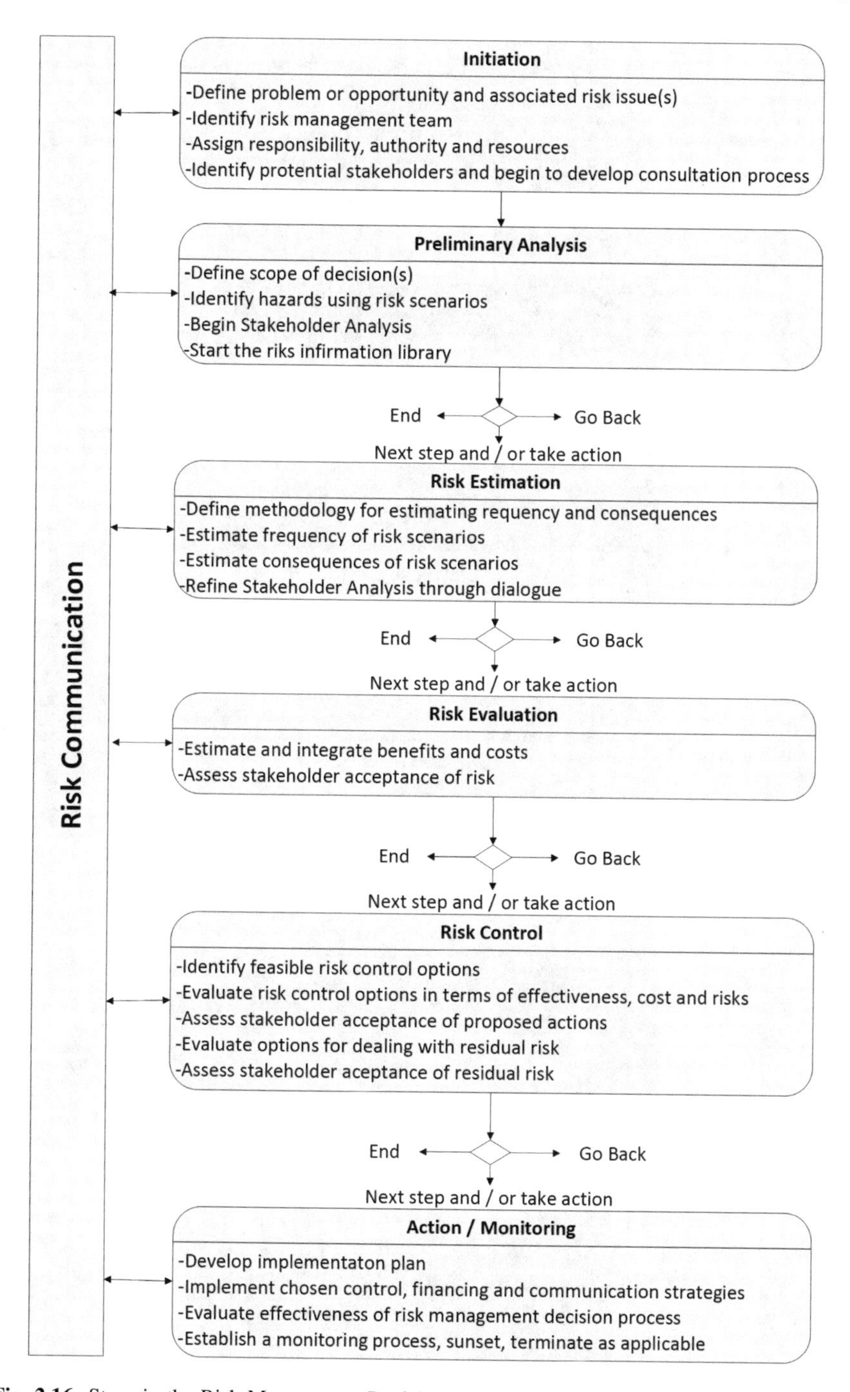

Fig. 2.16 Steps in the Risk Management Decision-Making Process—Detailed Model

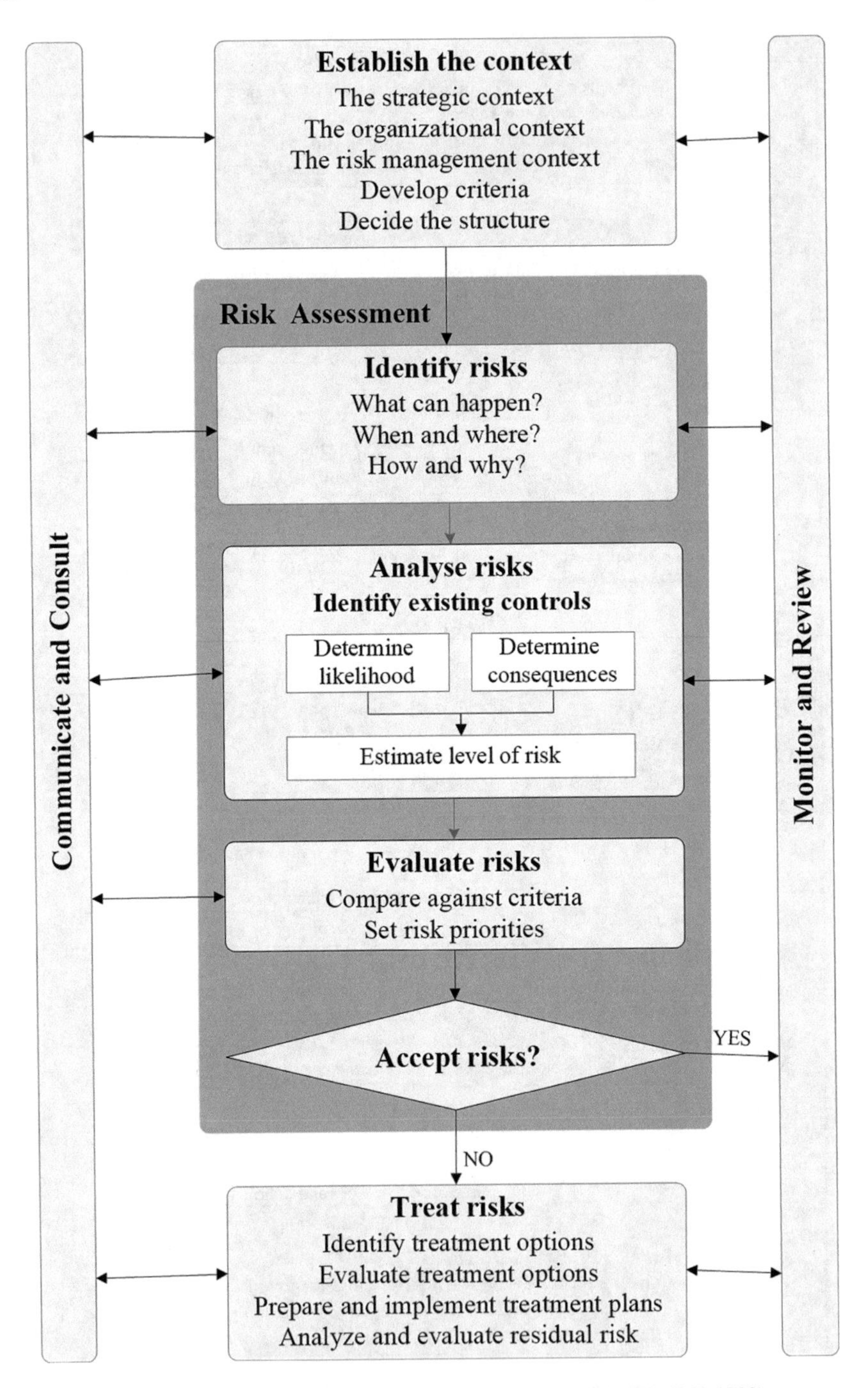

Fig. 2.17 Australia/New Zealand Risk Management Standards (AS/NZS 4360:1999)

The Australian-New Zealand framework introduced the concept of "Context" which is a further development of the Canadian "Initiation" step. Context provides the essential linkage between decision-makers and the technical or scientific analysis of risks. It is described as a process to "*Establish the strategic, organizational and risk management context in which the rest of the process will take place. Criteria against which risk will be evaluated should be established and the structure of the analysis defined.*"

The Australian framework also makes explicit the idea of Risk Criteria or "*terms of reference by which the significance of risks is assessed*" (ISO/IEC definition). These criteria can be quite broad and include qualitative and quantitative criteria, including absolute limits, social preferences, cultural, economic, and political criteria. The benchmark framework incorporates the Context and Risk Criteria concepts directly from this standard.

The Australian framework separates the assessment of the risk with already existing controls, and then if the remaining residual risk is not acceptable, additional treatment options are evaluated to make the residual risk acceptable.

The Australian-New Zealand standard also developed the concept of risk communication into "*communicate and consult*" and made it clear that any risk communication has to involve the decision-maker, an idea which was incorporated into the ISO/IEC definition and the benchmark framework.

According to the Standard, "*the success of risk management will depend on the effectiveness of the management framework providing the foundations and arrangements that will embed it throughout the Organization at all levels.*" Within the Standard, the expressions, "risk management" and "managing risks", are both used. In general terms:

- risk management refers collectively to the principles, framework and process for managing risks effectively, and
- managing risks refers to the application of these principles, framework and process to particular risks.

2.1.17 Comparison of Risk Management Frameworks

(Basic source: [19]).

At a high level, all described frameworks are similar in that each of them:

- sets out a generic process for risk management and accepts that there needs to be flexibility in implementation is applicable to a wide range of Organizations and activities
- recognises that management of risk is part of good management practice, should be continuous and is best embedded into existing practices/business processes.
- recognises that there can be positive outcomes, as well as negative outcomes,

- sets out steps in the risk management process with brief guidance on each.
- defines the terminology used.

A good comparison of the "*AIRMIC/ALARM/ IRM Risk Management Standard*" [20], the Australia/New Zealand Standard *AS/NZS 4360 [12]*, the *COSO Enterprise Risk Management - Integrated Framework* [9] is provided in [19].

Many of the principles of the standard ISO 31000 are similar to the risk management standard AS/NZS 4360 [5, 12]. However, ISO standardization gives a new definition of risk and 11 risk principles that are not presented in AS/ NZS 4360 (see Sects. 2.1.6 and 2.1.16).

The ISO 31000 standard is considered as a synthesis of best practices on risk management, referring to several previous Standards such as AS/NZS 4360 (Sect. 2.1.16) and COSO ERM (Sect. 2.1.10). The Australian and New Zealand standard, AS/ NZS, has been a major contributor to design the ISO standard 31000. Although the ISO 31000 standard has gained popularity in both Australia and New Zealand, this standard has not managed to replace AS/NZS standard completely [8].

2.1.18 The Effective and Robust Risk Framework

An effective and well implemented risk framework will [3] (Fig. 2.18):

- increase the likelihood of achieving Organization objectives
- improve the identification of opportunities and threats
- improve mandatory and voluntary reporting as well as overall governance
- comply with relevant legal and regulatory requirements
- effectively allocate and use resources for risk treatment
- maximise sustainable value
- enhance health and safety performance, as well as environmental protection
- improve Organizational learning and resilience
- improve loss prevention and incident management
- establish a reliable basis for decision making and planning
- improve Stakeholder confidence and trust.

2.2 Managing Changes

2.2.1 Introduction

The aim of the "change management process" is to ensure that changes are implemented in a safe manner to the extent that they are reasonably feasible. An effective "change management process" will also assist in consistent decision-making and ensure that the Infrastructure Manager or the Train Operating

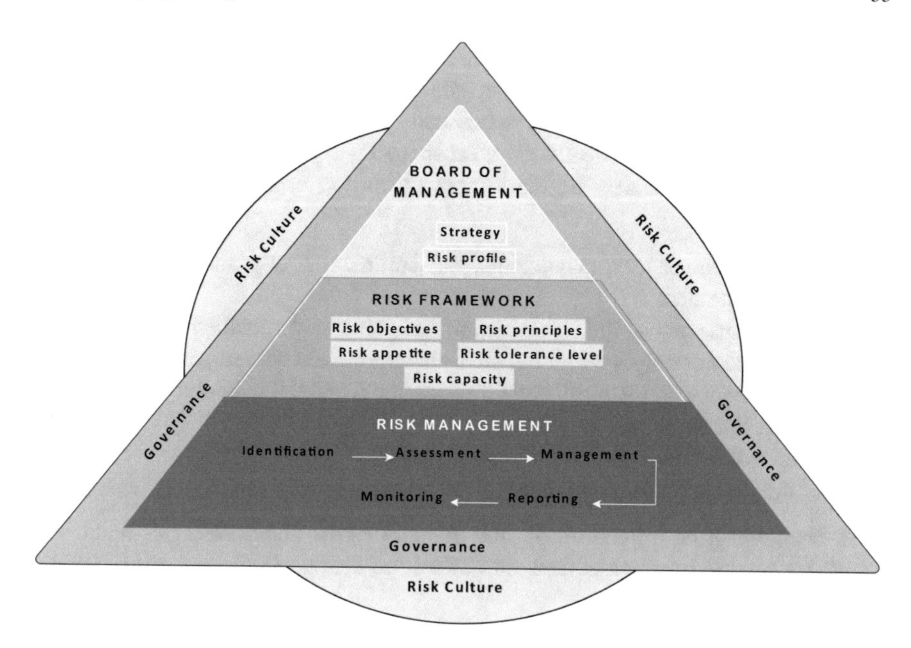

Fig. 2.18 The effective and robust risk framework (based on [3])

Company (Railway Undertaking) continues to comply with the legal framework and its certification conditions and restrictions.

Different types of change introduce different degrees of potential risk. The degree of the consideration required and the resulting level of detail at each step should be proportionate to the degree of risk that the change might introduce, or the process of implementing the change. Therefore, it is recommended that Infrastructure Managers or Train Operating Companies (Railway Undertakings) have several "change management processes" in place that requires increased analysis level as the potential level of risk associated with the change increases.

Large-scale changes, such as major infrastructure projects or Organizational restructuring, should be managed as a project with documentation on safety, such as documentation on safety validation, which is part of the change management plan. Throughout project planning, a change management plan should be treated as a live document and updated as information changes or become available. Similarly, the change management plan may initially set out the methodology and findings for risk assessment and later incorporate the requirements for safety.

2.2.2 A "Systems View" of Change

An Organization can be seen as a system of processes and people working together to achieve the Organization's purposes [21].

Changes inside systems frequently affect other parts of the system and can have unintended consequences unless the effects are thoroughly understood. The "change management process" is specifically intended to ensure that change's effects and influences are identified and managed.

Following elements may be the drivers of change or may be affected by the change:

a. People like employees, consultants, rail contractors, customers, suppliers and other Stakeholders
b. Environment, such as the Organization's physical and social environment. This may include not only the internal environment but also the surrounding industry (i.e. interfaces with other railway Organizations) and the regulatory environment in which the Organization exists.
c. Practices of work, policies and practices, and
d. Equipment, technology and equipment.

A change can be influenced by one or more of the above elements and/or the interfaces between these elements and affect the change in the risk profile of the overall Organization.

Thus, it is useful to consider the entire system (holistic view) in the change management process to ensure that the impact on all elements and their interfaces can be systematically identified, assessed and controlled throughout their lifecycle.

The Safety Management System of the Organization is a starting point for the process of defining the system and its relevant elements.

2.2.3 Change Management Procedures in a Safety Management System

The Safety Management System must include procedures for ensuring that changes that may affect the safety of railway operations are identified and managed, including but not limited to procedures for ensuring, so far as is reasonably practicable that:

1. the change is fully identified, described and documented in the context of the rail Organization
2. the changes are documented in a specific change register, the risk register or other appropriate means in the Safety Management System
3. affected parties are identified and, where practicable, consulted
4. the roles and responsibilities of staff with safety-critical positions and Employees of the Infrastructure Manager or the Train Operating Company are clearly specified with respect to the change
5. the risks to safety that may arise from the change are identified and assessed
6. the controls that are to be used to manage risks to safety and monitor safety are specified

7. the information in the risk register is updated with any changes to risks and control measures
8. the proposed change conforms to legislation
9. where appropriate, the change should also be consistent with accepted codes or standards
10. the staff with safety-critical positions and Employees of the Infrastructure Manager or the Train Operating Company are fully informed and trained to understand and deal with the proposed change
11. the change may involve a review of the competence requirements for the tasks to be undertaken
12. review and assessment of the change, once implemented is undertaken to determine whether the change has been appropriately managed
13. monitoring and review of the effect of the change should be undertaken, documented and necessary corrective actions implemented, to ensure that control measures perform as intended
14. decisions are transparent and formally accepted by those responsible for decision-making within the Infrastructure Manager or the Train Operating Company.

2.2.4 *Types of Change to Be Managed*

Changes to the Infrastructure Managers or Train Operating Companies (Railway Undertakings) can be from both internal and external sources.

Internal sources of change may include turnover in staff,[21] the findings or recommendations of internal audits, directions from the higher Management, findings from internal investigations, Organizational restructuring, or changes in the Organization's physical assets (such as new equipment).

External sources for change may include legislative or regulatory changes, Safety Investigation Authorities, Transport Authorities, other Infrastructure Managers or Train Operating Companies, suppliers or rail contractors.

The Infrastructure Manager's or Train Operating Company's change management procedures should enable the different types of change to be identified in advance and managed appropriately.

---

[21]Staff turnover refers to the number or percentage of workers who leave an organization and are replaced by new employees.

2.2.5 *Consultation During Change*

Consultation with persons affected by the change is an integral part of managing change and should be included, where reasonably practicable, at regular intervals throughout the "change management process". Proper consultation with key Stakeholders about proposed changes will ensure that risk is not transferred to those Stakeholders without their knowledge.

The level of consultation will be guided by the scale of change major change will likely require extensive and comprehensive consultation, whereas minor changes may only need a consultation with those directly managing the task.

The guiding principle is that it is better to consult with people no matter how little they may be affected by the change, as part of an open and transparent "change management process" and proper risk management. There are specific requirements for whom an Infrastructure Manager or Train Operating Company (Railway Undertaking) must consult with when establishing or reviewing a Safety Management System.

The objectives of the consultation are to:

a. exchange all information necessary for identification and assessment of the options for change and the possible impact of each option
b. ensure that all relevant railway personnel, interfacing external Organizations and other affected persons are aware of the proposed change, have an opportunity and are encouraged to comment on safety aspects and act consistently to achieve a safe outcome
c. ensure different perspectives are reflected in the monitoring and review of the proposed change and
d. promote ownership among affected persons for safety and the successful implementation of the change.

Consideration of the objectives of consultation assists the Infrastructure Manager or the Train Operating Company in identifying those who should be consulted in any case. Affected persons may include:

a. persons who will be involved in implementing the change (technical staff and/or end-user Employees staff with safety-critical positions) and/or whose work may be affected by the proposed change;
b. Organizations with an interface with the proposed change, maintenance and construction Contractors and other third parties whose access may be affected by changes to the scope of operations
c. manufacturers/suppliers
d. contractors
e. Unions representing affected persons
f. health and safety representatives under occupational health and safety (OHS) laws
g. the Public and/or local Communities.

It may also be beneficial to consult with persons who may make a useful contribution to the change process (for example, people with prior experience in similar changes, or who have technical expertise).

As a result of the consultation, the definition of the change, risk assessment, options and implementation plans may need to be amended.

Infrastructure Managers or Train Operating Companies should produce a Consultation Plan for changes involving multiple Stakeholders, or where a Stakeholder will be significantly affected by a change.

2.2.6 *Consultation Plans*

It may be useful to develop a Consultation Plan depending on the size of the project and the range of people to be consulted. Consultation Plans should be implemented with sufficient lead-time[22] to enable feedback and revision before implementation.

Elements of a Consultation Plan include but are not limited to the following:

a. name of the Infrastructure Manager or the Train Operating Company
b. title of the proposed change
c. brief description of the proposed change
d. Gantt chart or similar containing all elements of the plan including milestones, timelines and dates for achieving outcomes
e. names of all affected Infrastructure Managers or Train Operating Companies and other affected parties, including a brief description of the proposed consultation with each and their respective accountabilities
f. details of the communications packages to be prepared for all affected parties
g. details of all meetings, education/training of all affected personnel including a strategy that covers personnel not available during the implementation
h. means by which consultation responses, including risk and safety issues, will be received, addressed and fed into the project plan
i. resources required for consultation activities and
j. reporting requirements about the consultation undertaken.

2.2.7 *Steps in the "Change Management Process"*

An appropriate and robust change management and safety validation process involve seven main steps, which are set out in Fig. 2.19 [21].

[22]Lead time is the time between the original development of the consultation plan and its actual issue.

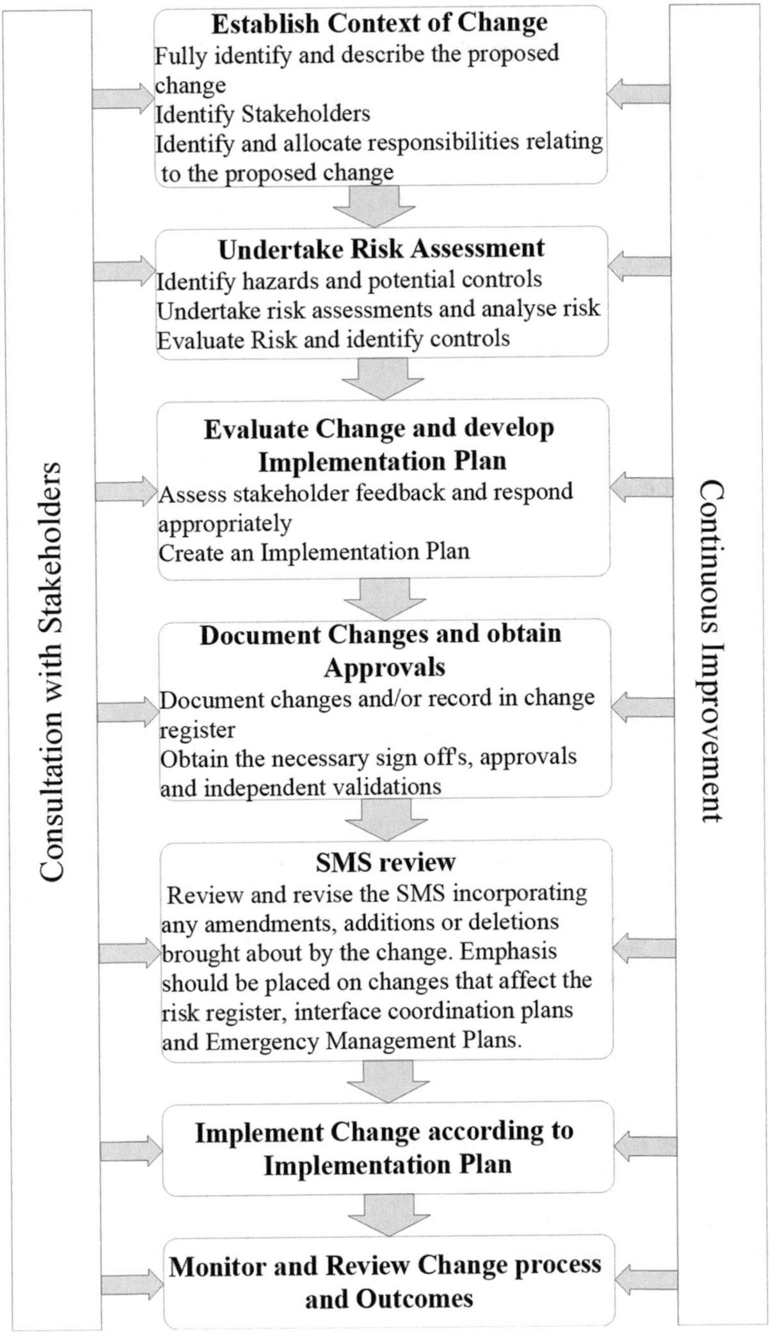

Fig. 2.19 Steps in the "change management process" (based on [21])

The amount of analysis and detail at each step should reflect the scale and risks involved with the change. For example, a riskier project would require more careful planning and risk analysis than a routine change.

There are two aspects of risk about a change:

- risks associated with the change itself and
- risks associated with the process of introducing the change.

It is a fundamental objective of the change management process that both aspects of risk are managed.

Step 1: Establish the Context of the Change and Consult with Stakeholders

1. This step involves identifying the change and developing the necessary plans for change management in consultation with Stakeholders, including interfacing Organizations.
2. A clear description of the current situation, including the problem or matter that the change seeks to address and the change itself, is required. This should be sufficiently detailed to define the overall nature and scope of the change fully. Changes can be defined and analysed at several levels, including project level, component level and/or process level. More than one may be applicable.
3. Where the Infrastructure Manager or the Train Operating Company has "change management processes" in place that require varying levels of analysis, the appropriate process is selected.
4. Each "change management process" should:

a. describe safety documentation requirements (such as safety validation documentation), including whether a change management plan is required
b. specify whether independent safety validation assessment is required and how that is to be achieved
c. identify the authority responsible for granting or refusing approval for implementation of the change (may include road authorities or other authorities outside of the Agency) and
d. provide criteria and guidance on the extent and nature of the consultation and meeting that should be carried out for the level of safety validation being applied.

Step 2: Undertake a Risk Assessment

1. This step is the actual undertaking of a risk assessment on the proposed change and deciding how the risk, including the controls, can be managed so far as is reasonably practicable. Appropriate use of risk management tools and techniques as part of the "change management process" ensures that the potential impacts are understood. This requires an in-depth understanding of the change proposed, its potential impacts on current activities, operational interfaces and the Infrastructure Manager's or the Train Operating Company's Safety Management System.

2. When an Infrastructure Manager or Train Operating Company undertakes a risk assessment, the emphasis is usually on any new incidents or associated hazards that could arise from the proposed change.
3. The legal framework (usually) requires that Infrastructure Managers or Train Operating Companies eliminate or reduce the risks to the safety of their operation so far as is reasonably practicable.
4. As a general principle, Infrastructure Managers or Train Operating Companies should be endeavouring to achieve a level of residual risk following the implementation of the change that is at least the same or better than the residual risk that existed before the implementation of the change.
5. Change may alter the balance of risk exposure to different groups. Infrastructure Managers or Train Operating Companies should endeavour to ensure an equitable balance of risk exposure to affected groups. Where the change involves a potential increase in risks to another party, the "change management process" should cover how those risks are likely to be increased and subsequently managed.[23]
6. Where existing risk controls are removed, the decision should be documented, explaining what controls have been removed and why and how the associated risks are to be managed.

Step 3: Evaluate Levels of Change and Develop an Implementation Plan

1. This step requires evaluation of the consolidated information gathered, further consultation (if practicable) with appropriate Stakeholders and making decisions on the options available. The change and associated activities are identified, and an implementation plan will be developed.
2. The implementation plan should address a range of matters, including:

    a. plans for introducing the change including all necessary modifications to the Safety Management System and Regulatory approvals
    b. communication, whereby significant changes regarding operations, equipment and procedures are effectively communicated throughout the Organization
    c. requirements for instruction and training
    d. any additional resources required to implement the change, for example, supervision or verification
    e. documents that need to be revised, for example, operating procedures, risk registers, training material, interface coordination plans, Emergency Management Plans and change management documentation itself and
    f. plans for monitoring and reviewing the change following implementation.

[23]For example, a change may reduce risk to a group, but introduce or increase risk to another group or an individual. In such circumstances there is a need to balance the risks affecting each group so that one group does not suffer very high levels of residual risk to reduce or remove the risk to the other.

Step 4: Document Changes and Obtain Approvals

1. This step involves consolidating documentation on the change, including any supporting records (such as external reports, etc.). The change should be clearly documented and signed from the appropriately authorised person or persons within the Infrastructure Manager or the Train Operating Company.
2. An independent safety validation where the proposed change relates to major projects should be undertaken by an appropriately experienced and/or qualified person who is sufficiently independent of the change.

Step 5: Review of the Safety Management System

This step involves the Infrastructure Manager or the Train Operating Company reviewing and revising where necessary, their Safety Management System, Risk Register, Emergency Management Plans and Interface Agreements.

Step 6: Implementation

1. Once a change has received the necessary internal and external approvals, the change may be implemented using the approved implementation plan.
2. The approved implementation plan must be fully carried out, including making all necessary modifications to Organizational documentation, such as the Safety Management System, risk assessments and other operational documentation.

Step 7: Monitoring and Review

1. The following questions should be asked at this step in the "change management process":

    a. have any new risks occurred, or pre-existing risks increased after implementation? Have any pre-existing risks been reduced or eliminated?
    b. are additional risk controls appropriate?
    c. have performance targets for the change been set and where applicable Organizational key safety performance targets been reviewed?
    d. has training been provided to staff affected by the change?
    e. has a post-implementation competency assessment been conducted to ensure the training provided was adequate for facilitating the change?
    f. is there a process to revise the risk assessment as new information accumulates?

2. Monitoring and review arrangements can be introduced immediately following the implementation of the change to ensure all risk controls, including training, have been effective and that documentation has been updated.

References

1. Shortreed J, Hicks J, Craig L (2003) Basic Frameworks for Risk Management, Final Report. Prepared for The Ontario Ministry of the Environment, Network for Environmental Risk Assessment and Management (NERAM)
2. Standards Australia/Standards New Zealand (2005) Risk management guidelines—companion to AS/NZS 4360:2004
3. Thornton G (2017) Risk frameworks—driving business strategy with effective risk frameworks
4. European Union Agency for Railways (2017) 'Safety Critical Components" in railways—the concept of "Safety Criticality" of the systems, Draft Report V0.1
5. ISO 31000, Risk management—principles and guidelines, 2009
6. PMI (Project Management Institute) (2009) Practice standard for project risk management, PMI
7. PMI (Project Management Institute) (2017) A guide to the project management body of knowledge (PMBOK® Guide), 6th edn. PMI
8. Remzi A, Besarta V (2017) Analysis of international risk management standards (advantages and disadvantages). Eur J Res Reflection Manag Sci 5(3)
9. Committee of Sponsoring Organizations of the Treadway Commission (COSO) (2004) Enterprise risk management—integrated framework
10. Intergovernmental Organization for International Carriage by Rail (OTIF) (2016) Uniform Technical Prescription—Common Safety Method on risk evaluation and assessment (UTP GEN-G consolidated version), 12 Jan 2016
11. The Chartered Institute of Management Accountants (CIMA) (2008) Enterprise risk management—Topic Gateway Series No. 49
12. Australian/New Zealand Standard, AS/NZS 4360 SET Risk Management, 2004
13. British Standard, BS 6079-3:2000/Project management. Guide to the management of business related project risk, 2000
14. The Institute of Risk Management, A Risk Management Standard (IRM/Alarm/AIRMIC) 2002
15. REGULATION (EU) No 402/2013 Common safety method for risk evaluation and assessment…, 2013
16. ORR (Office of Rail and Road), (UK), Common Safety Method for risk evaluation and assessment—Guidance on the application of Commission Regulation (EU) 402/2013, March 2015
17. The EU Agency for Railways—collection of examples of risk assessments and some possible tools: http://www.era.europa.eu/Document-Register/Documents/collection_of_RA_Ex_and_some_tools_for_CSM_V1.1.pdf
18. *CENELEC EN 50126-1:2017*—Railway Applications: The Specification and Demonstration of Reliability, Availability, Maintainability and Safety (RAMS). Generic RAMS Process
19. The Association of Insurance & Risk Managers (AIRMIC) (2005) An overview comparison of the AIRMIC/ALARM/ IRM Risk Management Standard with: the Australia/New Zealand Standard AS/NZS 4360:2004, the COSO Enterprise Risk Management—Integrated Framework
20. AIRMIC/ALARM/IRM Risk Management Standard, 2002 (Institute of Risk Management (IRM), The Association of Insurance and Risk Manager (AIRMIC) and The Public Risk Management Association (ALARM))
21. Office of the National Safety Regulator (Australia) (2013) Guideline for Preparation of a Safety Management System

22. Weeserik BP, Spruit M (2018) Improving operational risk management using business performance management technologies. Department of Information and Computing Sciences, Utrecht Organization, MDPI, Basel, Switzerland
23. Commission Regulation (EC) 352/2009: http://eur-lex.europa.eu/LexUriServ/LexUriServ.do?uri=OJ:L:2009:108:0004:0019:EN:PDF

Chapter 3
The Process of Risk Management

3.1 The Risk Management Processes

3.1.1 General

Risk management needs to be an integral part of proper management. It is an iterative process of continuous improvement in line with existing practices or business processes.

After examined the risk management framework and how it is applied in different countries in Europe, the American continent and Australia/New Zealand as also at different Organizations (Sect. 2.1.17), the risk management process as per the Australia/New Zealand standard [1] will be presented.

A brief overview of the risk management process is provided in the next section and all related steps will be discussed in greater detail in the next sections.

3.1.2 Main Elements of the Risk Management Process

The main elements of the risk management process, are the following (Fig. 3.1):

Establish the context	• Define the scope of enquiry/objectives: i.e. what activity, decision, project, program, issue requires analysis • Identify relevant Stakeholders/areas involved or impacted • Identify internal and/or external environmental factors
Identify risks	Identify/assess • What could happen? • How and where it could happen? • Why could it happen? • What is the impact or potential impact?

(continued)

K. Tzanakakis, *Managing Risks in the Railway System*, Springer Tracts on Transportation and Traffic 18, https://doi.org/10.1007/978-3-030-66266-0_3

(continued)

Analyse risks	• Identify the causes, contributing factors and actual or potential consequences • Identify existing or current controls • Assess the likelihood & impact/consequence to determine the risk rating
Evaluate risks	• Is the risk acceptable or unacceptable? • Does the risk need treatment or further action? • Do the opportunities outweigh the threats?
Treat risks	• If existing controls are inadequate, identify further treatment options • Develop a treatment plan • Seek endorsement & support for treatment • Determine the residual risk rating once the risk is treated
Monitor and review	• Continually check –Effectiveness of risk controls and/or treatments –Changes in context or circumstances, and • Document & report this activity accordingly
Communicate and consult	At all stages of the process • Ensure those responsible for managing risk, and those with vested interests, understand the basis on which decisions are made, why particular treatment options are selected, or why risks are accepted/ tolerated

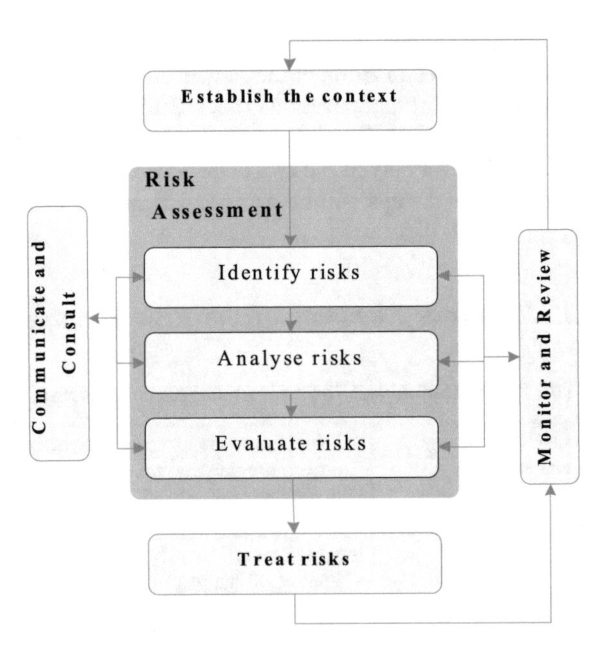

Fig. 3.1 The risk management processes—an overview

Risk management in an Organization can be applied at many levels. It can be applied at a strategic, tactical and operational level. It can be applied to specific projects, supporting specific decisions or managing specific recognized areas of risk.

It is necessary to keep records for each stage of the process so that decisions can be understood as part of a process of continuous improvement.

The details of the risk management process are shown in Fig. 3.2 and explained in the next sections.

3.2 Establish the Context

3.2.1 The Business Objectives and the Environment of the Organization

Establishing the context has to do with understanding the background of the Organization and its risks. It aims to provide a comprehensive consideration of all the factors that may influence the ability of an Organization to achieve its intended outcomes. The output from this step is a brief statement of the organizational objectives and the setup of the specific criteria for success, the objectives and scope for risk management. The context must be clearly defined so that the rest of the process stays within the required limitations.

Objectives and the external and internal environment
As discussed, risk is the chance of something happening that will have an impact on objectives. Therefore, we must ensure that all significant risks are captured.

The identification of the external and internal environment is the first step in establishing the context. For this, key documents, such as the strategic plan, business plans and budgets, annual reports, economic analyses, and any other relevant documentation about the Organization and its purpose may be consulted. External documents such as relevant legislation should also be consulted. Strategic analysis documents, such as SWOT analyses (Sect. 4.1.9.3.4), may be valuable, as they assist in focussing on relevant aspects of the external and internal environment.

Establish the internal context (internal environment)
The internal context is the internal environment in which the Organization functions and seeks to achieve its objectives. Consideration should be given to factors such as:

- Objectives and strategies in place to achieve objectives
- Governance, structure, roles and accountabilities
- The capability of people, systems and processes
- Changes to processes or compliance obligations
- The risk tolerance and appetite of the Organization.

Establish the external context (external environment)
The external context is the environment in which the Organization operates and seeks to achieve its objectives. Consideration should be given to the following

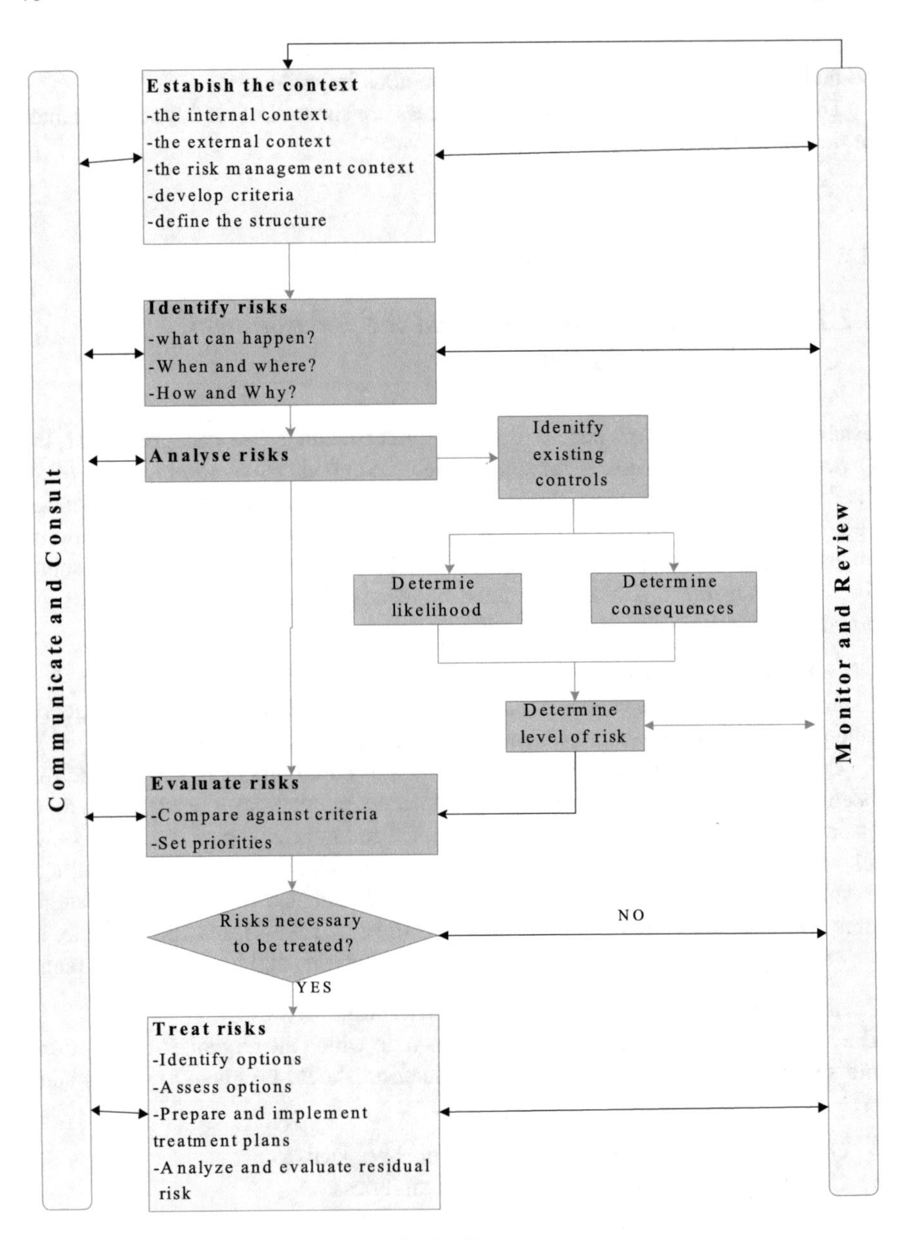

Fig. 3.2 The risk management processes in detail

inputs as they relate to the business; the social, regulatory, legislative, cultural, competitive, financial, and political environment, including:

- Strengths, weaknesses, opportunities and threats
- Relationships with, perceptions and values of, external Stakeholders such as clients.

3.2.2 *The Stakeholder of the Organization*

Stakeholder identification and analysis are important in risk management. It is usually undertaken at an early stage of planning.

Stakeholders are those who may affect, be affected by, or perceive themselves to be affected by the Organization or the risk management process. In other words, Stakeholders are those people or groups who have a legitimate interest in the Organization.

A simplified presentation of the Stakeholders is provided in Fig. 3.3.

The objectives of Stakeholder management are to provide a clearer understanding of the Stakeholders and, as a result, provide insights as to how best to engage them (Fig. 3.4).

Stakeholder management has three stages:

- Identification
- Analysis
- Communication

Stakeholder Identification The first step in the Stakeholder analysis is to brainstorm who are the Stakeholders. As part of this, think of all the people who are affected by the work, who have influence or power over it, or have an interest in its successful or unsuccessful conclusion. Remember that although Stakeholders maybe both Organizations and people, ultimately with people. Make sure that the correct individual Stakeholders within a Stakeholder Organization are identified.

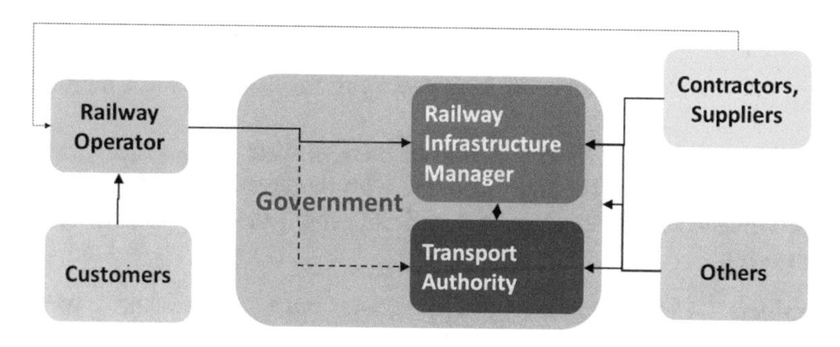

Fig. 3.3 Stakeholders of the railway sector (simplified)

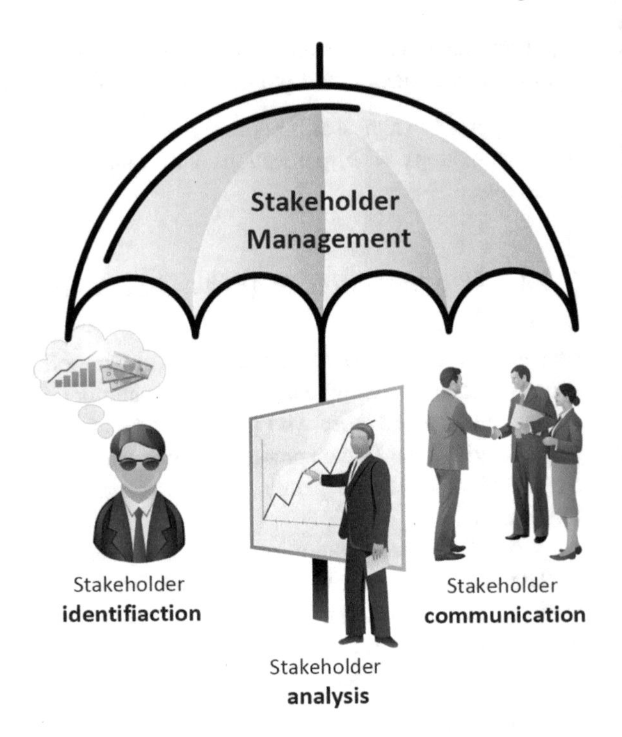

Fig. 3.4 Stakeholder management

Stakeholders Analysis: Stakeholder analysis provides decision-makers with a documented profile of Stakeholders to better understand their needs and concerns.

Stakeholders should be appropriately engaged at each point and cycle of the risk management process through a process of communication and consultation. Involving Stakeholders can build acceptance and be constructive.

The main aims and objectives of relevant Stakeholders should be considered explicitly. This may take a quite simple form, or more sophisticated analyses may be appropriate where major social and community risks are anticipated. Stakeholder analysis matrix can help in the analysis process. A Stakeholder's position in a quadrant suggests an approach to be considered (Fig. 3.5):

- High influence/high interest = fully involve and make the greatest efforts to satisfy them
- Low influence/high interest = keep them informed and request their input to relevant areas
- High influence/lower interest = consult them in their interest areas, but not so extensively that they become frustrated with the level of detail
- Low influence/low interest = monitor these people and keep them informed at a high level.

Stakeholders Communication Stakeholders must understand what the Organization is trying to achieve. Stakeholders Communication builds an understanding of the Organization's goals and the benefits to the audience if they help

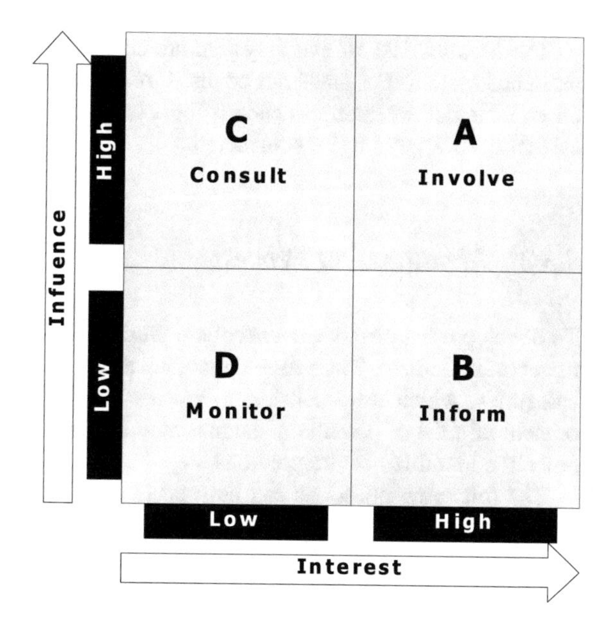

Fig. 3.5 Stakeholder analysis matrix

you achieve those goals. Communication helps the Organization to build positive relationships with people and other Organizations, such as the media or special interest groups, who influence other Stakeholders. The communication program must focus on the Stakeholders who have the greatest influence on the Organization's success.

3.3 Identify Risks

Identify what can happen, where and when it can happen

3.3.1 *Aim*

Risk identification has to develop a list of sources of risks and events that might have an impact on the achievement of each of the objectives identified in the previous step: establish the context.

The list should include all risks that can be a (significant) threat to our Organization or have as a consequence that significant opportunities are being missed.

The hazards/risks of every system are covered either by a code of practice or a reference system. For areas not covered by codes of practice or reference systems, an explicit risk evaluation needs to be executed (Fig. 3.6). The interfaces of the system need always to be considered.

3.3.2 *Identification Process*

To develop a list of risks, a systematic process should be used that starts based on the context defined. The process includes a review of the key Organizational risk categories, which were considered when establishing the context, and the development of a list of potential risks that may impact the Organization achieving each objective identified in the previous step.

The following questions can be used to assist in identifying risks:

- What could go wrong?
- How could we fail?

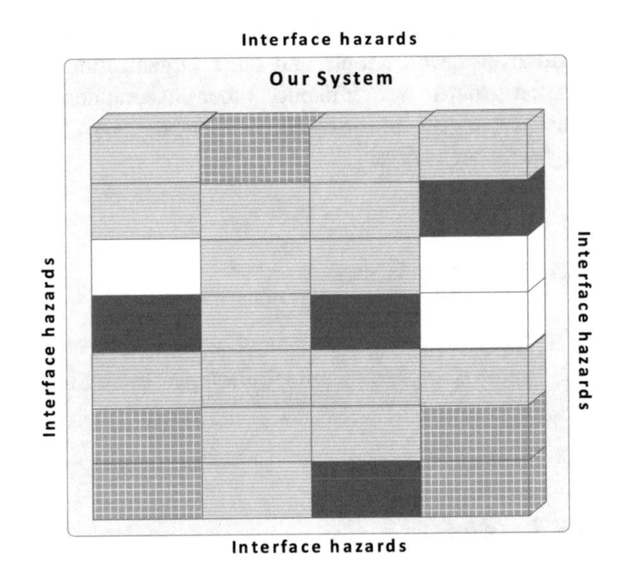

Fig. 3.6 Hazard identification and acceptance

- What must go right for us to succeed?
- Where are we vulnerable?
- What assets do we need to protect?
- Do we have liquid assets or assets with alternative uses?
- How could someone steal from the Organization?
- How could someone disrupt our operations?
- How do we know whether we are achieving our objectives?
- On what information do we most rely on?
- On what do we spend the most money?
- How do we bill and collect our revenue?
- What decisions require the most judgment?
- What activities are most complex?

3.3.3 Methods for Identifying Risks

Good quality information is crucial in the identification of risks. Historical records are the starting point for risk identification; some examples include:

- Expert judgment.
- Structured interviews.
- Focus group discussions.
- Strategic and business plans, including SWOT analysis and environmental scanning.
- Insurance claims reports.
- Post-event reports.
- Personal experience or past organizational experience.
- Results and reports from audits, inspections and site visits.
- Surveys and questionnaires.
- Checklists.
- Historical records, incident databases and analysis of failures and previous risk registers if they exist.

People involved in identifying risks must be well-informed about the detailed aspects of the risk study being undertaken. Teams allow sharing experiences, building engagement and ownership into the process of risk management.

3.3.4 Documentation of the Risk Identification Process

Documentation of the risk identification process should include

a. the approach or method used
b. the scope covered by the identification

c. the participants in the risk identification and the information sources consulted and
d. a risk register (see Sect. 4.2.2).

3.4 Risk Analysis

3.4.1 Introduction to the Risk Analysis

3.4.1.1 The Aim and the Main Steps for the Risk Analysis

Risk analysis aims to

- establish an understanding of the level of risk and its nature and
- help to set treatment priorities and options.

Risk consists of two parts: the likelihood of something going wrong, and if it does, the negative consequences. The level of risk is defined by combining consequence and likelihood, as we will discuss in detail in the next sections.

Having identified the risks and described the context, causes, contributing factors and consequences, we must look at the strengths and weaknesses of existing systems and processes designed to help control the risk. Knowing what controls are already in place and whether they are effective, helps to determine what further action is needed if any.

The risk analysis process has several steps:

Step	Question	Comment
1. Identify the controls to mitigate the impact of risks	What are the existing controls?	Identify existing controls–determine which controls are in place to mitigate the risk impact. Controls can be strong or weak; measurable and reproducible. Controls may include legislation, policies or procedures, training of staff, segregation of duties, personal protections and equipment, and structural or physical barriers (i.e. setting up IT firewalls) (see Sect. 3.7—Risk treatment)
Once the controls have been identified, and their effectiveness analysed, an evaluation is made of the likelihood of the risk that may occur and the consequence if the risk occurs. This produces an accurate, though subjective, risk level assessment -or risk rating- and helps determine whether risks are acceptable or need further treatment in the next step		

(continued)

(continued)

2. Assess the likelihood of risks	How likely is the risks identified to occur?	the likelihood of the risk occurring is described as Rare, Unlikely, Moderate, Probable and almost certain to occur (Sect. 3.4.4)
3. Assess the consequence	What are the consequences if the risk were to occur?	the consequences or potential impact if the risk event occurred can be described as negligible, minor, moderate, major, severe (Sect. 3.4.3)
The assessment of likelihood and consequence is mostly subjective but can be informed by collected data or information, audits, inspections, personal experience, corporate knowledge or institutional memory of previous events, insurance claims, surveys, and a variety of other internal and external information available		
4. Rate the level of risk	How high are the risks?	use the Risk Matrix of your Organization to assess the likelihood and consequence levels; the risk matrix then determines whether the risk rating is low, medium, high or extreme. The Risk Matrix also identifies the management action required for the various risk ratings

The process of risk analysis will often start with a simple qualitative approach that gives a general understanding. Where greater detail or understanding is required, a more focused and robust investigation may be needed as well.

3.4.1.2 Principles for the Estimation of Risk (Risk Analysis)

Risk is the combination of its two components, namely, consequence and likelihood. The relationship between these two components will depend on many factors as per the nature of the risk and the way it is perceived.

In its simplest form risk can be shown as:

$$Risk = f(Consequence, Likelihood) \tag{3.1}$$

If the level of risk is proportional to each of its two components (consequence or likelihood), the risk function can be represented as:

$$\boxed{Risk = C * L} \tag{3.2}$$

where

C = Consequence, L = Likelihood

This simple relationship, as shown in formula (3.1) does not consider complicating factors such as nonlinear relationships between risks and consequences. So, for quantitative analysis, we may need to include a weighting factor for one of the two components (to achieve a required relative scale between them) and may also require an exponential operator[1] for one or both components [2]. For example:

$$Risk = (C * w)^x * L^y \tag{3.3}$$

where
w = weighting factor

3.4.1.3 Key Questions in Analysing Risk

The following are ten questions for management and boards to consider[2]:

1. What are the Organization's top risks, how severe is their impact, and how likely are they to occur?

   Managing Organization risk at a strategic level requires focus, meaning generally emphasizing no more than five to 10 risks. Day-to-day risks are an ongoing operating responsibility.

2. How often does the Organization refresh its assessment of the top risks?

   The risk assessment process of the Organization should be responsive to change in the business environment. A robust process for identifying and prioritizing the critical Organization's risks, including emerging risks, is vital to an evergreen view of the top risks.

3. Who owns the top risks and is accountable for results, and to whom do they report?

   Once the key risks are identified, someone or team, or Organization unit must own them. Gaps and overlaps in risk ownership should be minimized, if not eliminated.

4. How effective is the Organization in managing its top risks?

   A robust process for managing and monitoring each of the critical Organization's risks is essential to successful risk management, and risk management capabilities must be improved continuously as the speed and complexity of business change.

5. Are there any organizational "blind spots" warranting attention?

   Cultural issues and dysfunctional behaviour can undermine the effectiveness of risk management and lead to inappropriate risk-taking or the undermining of established

---

[1]raise to power' operators, x and y for C and L.

[2]See https://www.corporatecomplianceinsights.com/ten-questions-you-should-ask-about-risk-management/

policies and processes. For example, lack of transparency, conflicts of interest, a shoot-the-messenger environment and/or unbalanced compensation structures may encourage undesirable behaviour and compromise the effectiveness of risk management.

6. Does the Organization understand the key assumptions underlying its strategy and align its competitive intelligence process with monitoring external factors for changes that could alter those assumptions?

An Organization can fall so in love with its business model and strategy that it fails to recognize changing models until it is too late. While no one knows for sure what will happen that could invalidate the Organization's strategic assumptions in the future, monitoring the validity of key assumptions over time as the business environment changes is a smart thing to do.

7. Does the Organization articulate its risk appetite and define risk tolerances for use in managing the business?

The risk appetite dialogue helps to bring balance to the conversation around which risks the Organization should take, which risks it should avoid and the parameters within which it should operate going forward. The risk appetite statement is decomposed into risk tolerances to address the question, "How much variability are we willing to accept as we pursue a given business objective?" For example, separate risk tolerances may be expressed differently for objectives relating to earnings variability, interest rate exposure, and the acquisition, development and retention of people.

8. Does the Organization's risk reporting system provide to the top management information they need about the top risks and how they are managed?

Risk reporting starts with relevant information about the critical Organization's risks and how those risks are managed. Are there opportunities to enhance the risk reporting process to make it more effective and efficient? Is there a process for monitoring and reporting critical Organization's risks and emerging risks to executive management and the board?

9. Is the Organization prepared to respond to extreme events?

Does the Organization have response plans for unlikely extreme events? Has it prioritized its high-impact, low-likelihood risks in terms of their reputational effect, velocity to impact and persistence of impact, as well as the Organization's response readiness?

10. Do the high-level managers have the requisite skill sets to provide effective risk oversight?

To provide input to executive management regarding critical risk issues on a timely basis, directors must understand the business and industry, as well as how the changing environment impacts the business model.

3.4.2 *Qualitative and Quantitative Risk Analysis*

3.4.2.1 Introduction

The basis of risk management is the expertise on how to select the right risk analysis methods. The choice of the proper analysis method depends on the context, objectives and available resources of the Organization. At a **strategic level**, broad categories of risk may be identified and analysed to provide an organizational risk profile, showing important issues for which specific management systems and risk treatments need to be established. At a **team level**, managers need to identify and prioritize the specific risks that threaten the objectives they are assigned to them and they need to achieve.

There are two methods for the risk analysis, the qualitative and quantitative analysis method. Superior to qualitative analysis is the quantitative risk analysis.

- **Qualitative risk analysis** prioritises the risks using a pre-defined rating scale. Risks will be scored on their probability of happening.
- **Quantitative risk analysis** is a further level of risk analysis of the highest priority threats and is usually assigned a numerical value i.e. cost of risk to the business and/or project tasks.

In the next Sections, the two methods for the risk analysis are discussed in detail.

Some risks may need to be examined in more detail. Reasons for detailed analysis, which may be quantitative or qualitative, are [3]:

a. to obtain more information about consequences or likelihood so decisions about priorities are based on information and data rather than guesswork
b. to better understand the risk and its causes so that treatment plans can be directed at true rather than superficial causes of problems
c. where decision criteria require more in-depth analysis (often this is where decision criteria are expressed quantitatively)
d. to help people choose between options where each has different costs and benefits and potential opportunities and threats
e. to provide a better understanding of risk to individuals who must operate with the risks or
f. to provide an understanding of residual risk after treatment strategies have been applied.

3.4.2.2 Qualitative Risk Analysis

Qualitative analysis is any method of analysis which uses description rather than numerical means to define a level of risk. As mentioned, qualitative risk analysis prioritises the risks using a pre-defined rating scale. Risks will be scored on their probability of happening.

The qualitative risk analysis method can be applied to assist in deciding how serious a risk is. It should be applied to all risks identified as requiring assessment including risks associated with change.

- Risks that give a low rating in this screening can be monitored or accepted.
- Those with a medium rating must be reduced at a reasonable cost.
- Those with a high rating need immediate action to reduce the risk.

Qualitative analysis may be used

a. where quantitative precision is not needed
b. to perform an initial screening of risks prior to further, more detailed analysis
c. where the level of risk does not justify the time and resources needed to do a numerical analysis or
d. where the numerical data are not available or inadequate for more quantitative analysis.

3.4.2.3 Quantitative Risk Analysis

The level of risk can be calculated using a quantitative method when the consequences and likelihood of occurrence can be quantified. For example, train collision risk assessments may be quantitative, where the likelihood can be expressed numerically, and the potential consequences (impacts) can be measured in terms of monetary loss. So, rather than ranking the likelihood of a risk as a 2 on a scale of 1–5 qualitatively (see Sect. 3.4.4), we would estimate the probability to be 5%. Rather than ranking the consequence (impact) as a 3, we would say that the impact is three weeks or $200,000.

3.4.2.4 The Difference Between Qualitative and Quantitative Risk Analysis

When should you perform quantitative risk analysis? You perform quantitative risk analysis when you need to quantify the risks and understand the risks at a deeper level. Think of a physical health exam. The doctor may ask you questions (qualitative analysis), but he or she may also choose to do laboratory tests (quantitative analysis) to get a deeper understanding of what's going on.

Next, the main differences between qualitative and quantitative risk analysis are presented.

Qualitative risk analysis	Quantitative risk analysis
• Should always be performed • Subjective • Quick	• Optional • Objective/Numeric • Takes more time • Provides more in-depth information i.e. about the probability of completing the project on schedule and within budget • Need to develop a contingency reserve

3.4.3 Estimation of the Consequences

To estimate the consequences of a risk, we can use tables that are consistent with the specific objectives and the context of a risk management activity.

Next, we will present three examples of simple qualitative consequence tables that might be used by an Organization:

1. a very simple descriptive Table 3.1
2. a simple descriptive table with criteria related to financial damages and effect on people (Table 3.2)
3. a simple descriptive table with criteria related to health and safety, environment and financial success as also political and financial impacts of risks (such as might be come across in the public sector) (Table 3.3).

3.4.4 Estimation of the Likelihood of Risk (Hazard)

To estimate the likelihood of a risk if, we need to use tables to meet the circumstances of each case.

Next, we show three examples of likelihood tables (scales) that might be used by an Organization. As per the tables, we can select the likelihood rating of the risk (hazard):

Table 3.1 Simple consequence scale (example)

Rating	Descriptive	Definition
1	Negligible	Negligible impact upon objectives
2	Minor	Minor effects that are easily remedied
3	Moderate	Some objectives affected
4	Major	Some important objectives cannot be achieved
5	Severe	Most objectives cannot be achieved

Table 3.2 Simple consequence scale with criteria related to financial damages and effect on people (example)

Rating	Description	Financial damage (example, USD)	Effect on people
1	Negligible	<10,000	No medical treatment by professional medical personnel
2	Minor	10,000–100,000	Little or no injuries and isolated damage
3	Major	100,000–1 million	Limited number of fatalities/casualties and damage to property
4	Critical	1 million–10 million	Numerous fatalities/casualties, loss of essential services, and widespread damage
5	Catastrophic	>10 million	More than one fatality (multi-fatality) and/or multiple severe injuries

1. Table 3.4 is an example showing a scale that is more suited to a defined period of time where the absolute likelihood of an event may be related to given activities—a project for example where the chance of achieving a certain outcome may need to be considered.
2. Table 3.5 uses order of magnitude scales to span a range of likelihoods from approximately yearly to one in 10,000 years.

Systems engineering techniques such as fault tree analysis can be used to analyse probabilities in more detail (see Sect. 4.1).

3.4.5 *Estimation of the Level of Risk*

The way that the level of risk is described will depend upon the type of analysis undertaken. **Qualitative analysis** can only describe risk in qualitative ways, usually done with descriptive terms (low, medium, high risk) (see Sect. 3.4.2). An example of this is given in Tables 3.6 and 3.10.

On the other hand, **quantitative analysis** (Sect. 3.4.2) may produce a single figure, datum or value or a mass of detailed data. Care must be taken when applying quantitative analysis when we are examining consequences that are intangible or difficult to quantify such as environmental or safety effects or reputation.

The categories we define may be linked to the level of management attention or response required. For example:

a. **Low risk** to be managed by routine procedures, specific application of resources (management attention is not likely to be required).
b. **Medium risk** can be managed by specific monitoring or response procedures (management responsibilities to be specified).
c. **Very high or high risk**: senior executive management attention is needed, action plans and management responsibility need to be specified.

Table 3.3 Consequences scale with criteria related to health and safety, the environment, financial success, political and financial impacts (as per [2])

Rating (severity level)	Consequences types					
	Profit Reduction (USD)	Health and safety	Natural environment	Social/cultural heritage	Community, Government, Reputation, Media	Legal
1	<10,000	No medical treatment required	Minor effects on biological of physical environment	Minor medium-term social impacts on local population Mostly repairable	Minor, adverse local public or medical attention or complaints	
2	10,000–100,000	Objective but reversible disability requiring hospitalization	Moderate, short-term effects but not affecting ecosystem functions	On-going social issues. Permanent damage to items of cultural significance	Attention from media and/or heightened concern by local community. Criticism by NGOs	Minor legal issues, non-compliances and breaches or regulation
3	100,000–1Million	Moderate irreversible disability or impairment (<30%) to one or more persons	Serious medium term environmental effects	On-going serious social issues. Significant damage to Structures/ items of cultural significance	Significant adverse national media/public/ NGO attention	Serious breach of regulation with Investigation or report to authority with prosecution and/ or moderate find possible
4	1 Million–10 Million	Single fatality and/or severe irreversible disability (>30%) to one or more persons	Very serious, long-term Environ-mental impairment of ecosystem functions		Serious public or media outcry (international coverage)	Major breach of regulation. Major litigation
5	10 Million–100 Million	Multiple fatalities, or significant irreversible effects to >50 persons				Significant prosecution and fines. Very serious litigation including class actions
IV	1 Million–10 Million	Single fatality and/or severe irreversible disability (>30%) to one or more persons		Ongoing serious social issues. Significant damage to structures/ items of cultural significance	Serious public or media outcry (international coverage)	Major breach of regulation. Major litigation

Table 3.4 Example likelihood scale (probability)—Example 1

Level	Description	Definition
1	Rare	Unlikely to occur but possible. It can be assumed the risk (hazard) may exceptionally occur
2	Unlikely	It can be reasonably expected for the risk (hazard) to occur
3	Moderate	Highly possible for the risk (hazard) to occur
4	Probable	(Almost certain) The risk (hazard) can be expected to occur frequently
5	Almost certain	Risk (hazard) is certain to occur or already has

Table 3.5 Example likelihood scale—Example 2 (based on [2])

Level	Descriptor	Description	Indicative frequency (expected to occur)
1.	Almost certain	The event will occur on an annual basis	Once a year or more frequently
2.	Likely	The event has occurred several times or more in your career	Once every three years
3.	Possible	The event might occur once in your career	Once every ten years
4.	Unlikely	The event does occur somewhere from time to time	Once every thirty years
5.	Rare	Heard of something like the occurring elsewhere	Once every 100 years
6.	Very rare	Have never heard of this happening	One in 1000 years
7.	Almost incredible	Theoretically possible but not expected to occur	One in 10 000 years

3.4.5.1 Estimation of the Level of Risk as Per the "Consequences and Likelihood" Matrix

After examined the consequences and likelihood of a risk (hazard) we can assess the risk using a 5 × 5 **risk matrix** es shown below. Risk can be calculated by multiplying the likelihood rating and the consequence rating:

Risk = Likelihood * Consequence

Table 3.6 The "Consequences—Likelihood" matrix

			Consequence				
			Negligible	Minor	Major	Critical	Cata-strophic
		Ra-ting	1	2	3	4	5
Likelihood	Rare	1	1	2	3	4	5
	Unlikely	2	2	4	6	8	10
	Moderate	3	3	6	9	12	15
	Probable	4	4	8	12	16	20
	Almost certain	5	5	10	15	20	25

Where

Key	Pattern	Risk Rating	Action
Low		0-9	Risk can be monitored or accepted.
Medium		10-16	Risk should be reduced at reasonable cost. (Apply ALARP principle, see Section 16.4)
High		20-25	Immediate action required to reduce risk

3.4.5.2 Estimation of the Level of Risk as Per the "Frequency-Consequences" Matrix

This section deals with the formation of a "**frequency-consequence**" **matrix** for the evaluation of the results of risk analysis, the risk categorisation, actions required for risk reduction or elimination of intolerable risks, and for risk acceptance.

Risk evaluation shall be performed by combining the frequency of occurrence of a hazardous event with the severity of its consequence to establish the level of risk generated by the hazardous event. A "frequencys-consequence" matrix is shown in Table 3.7 (after EN 50126.1).

Table 3.8 provides, in qualitative terms, typical categories of probability or frequency of occurrence of a hazardous event and a description of each category for a railway system. The categories, their values, and their numerical scaling to be applied shall be defined by the Infrastructure Manager, appropriate to the application under consideration.

Consequence analysis shall be used to estimate the likely impact.

Table 3.9 describes typical **hazard severity levels** and the consequences associated with each severity level for all railway systems. The number of severity

Table 3.7 The "Frequency-Consequence" matrix

Frequency of occurrence of a hazardous event	Risk levels			
	Severity levels of hazard consequence			
	Insignificant	Marginal	Critical	Catastrophic
Frequent				
Probable				
Occasional				
Rare				
Improbable				
Incredible				

Table 3.8 Frequency of occurrence of hazardous events (after EN 50126.1)

Category	Description
Frequent	Likely to occur frequency. The hazard will be continually experienced
Probable	Will occur several times. The hazard can be expected to occur often
Occasional	Likely to occur several times. The hazard can be expected to occur several times
Rare	Likely to occur sometime in the system life cycle. The hazard can reasonably be expected to occur
Improbable	Unlikely to occur but possible. It can be assumed that the hazard may exceptionally occur
Incredible	Extremely unlikely to occur. It can be assumed that the hazard may not occur

Table 3.9 Hazard Severity Level (after EN 50126.1)

Severity Level	Consequence to persons or environment	Consequence to service
Insignificant	Possible minor injury	Minor system damage
Marginal	Minor injury and/or significant threat to the environment	Severe system(s) damage
Critical	Single fatality and/or severe injury and/or significant damage to the environment	Loss of a major system
Catastrophic	Fatalities and/or multiple severe injuries and/or major damage to the environment	

levels and the consequences for each severity level shall be defined by the Infrastructure Manager.

The tools for the hazard/risk identification are discussed in detail in Sect. 4.1.

Table 3.10 shows an example of risk evaluation and risk reduction/controls for risk acceptance.

Table 3.11 defines qualitative **categories of risk** and the actions to be applied against each category. The Organization shall be responsible for defining principle

Table 3.10 Typical example of risk evaluation and acceptance (after EN 50126.1)

Frequency of occurrence of a hazardous event (Scaling for the frequency of occurrence of hazardous events will depend on the application under consideration)	Risk levels			
	Severity levels of hazard consequence			
	Insignificant	Marginal	Critical	Catastrophic
Frequent	Undesirable	Intolerable	Intolerable	Intolerable
Probable	Tolerable	Undesirable	Intolerable	Intolerable
Occasional	Tolerable	Undesirable	Undesirable	Intolerable
Rare	Negligible	Tolerable	Undesirable	Undesirable
Improbable	Negligible	Negligible	Tolerable	Tolerable
Incredible	Negligible	Negligible	Negligible	Negligible

Table 3.11 Qualitative Risk Categories (after EN 50126.1)

Risk evaluation	Risk reduction/control
Intolerable	Shall be eliminated
Undesirable	Shall only be accepted when risk reduction is impracticable and with the agreement of the Infrastructure Manager
Tolerable	Acceptable with adequate control and the agreement of the Infrastructure Manager
Negligible	Acceptable without any agreement

to be adopted and the tolerability level of risk and the levels that fall into the different risk categories.

3.4.6 Performance and Documentation of Risk Analysis

Risk analysis shall be performed at various phases of the system life cycle by the Organization responsible for that phase and shall be documented. The documentation shall contain, at a minimum:

a. analysis methodology
b. assumptions, limitations and justification of the methodology
c. hazard identification results
d. risk estimation results and their confidence levels
e. results of trade-off studies
f. data, their sources and confidence levels
g. references.

3.4.7 Risk as Opportunities (Positive Risks)

The entire practice of risk management is focused on controlling potential negative outcomes. However, the risk management can be used to identify and prioritize opportunities (or "positive" risks) following almost the same process as presented in previous sections.

The likelihood of opportunities should be related to the nature of the expected beneficial outcomes.

Table 3.12 is showing an example of describing the positive consequences that correspond to the needs and nature of an Organization. This table is similar to the negative consequence tables presented in Sect. 3.4.3.

Table 3.12 Example of detailed description for positive consequence (see [2])

Level	Descriptor	Description
1	Insignificant	Small benefit, low financial gain
2	Minor	Minor improvement to image, some financial gain
3	Moderate	Some enhancement to reputation, high financial gain
4	Major	Enhanced reputation, major financial gain
5	Outstanding	Significantly enhanced reputation, substantial financial gain

By combining the likelihood and consequence ratings, we can determine the level of opportunity:

• Very high opportunity	Detailed planning required at senior levels to prepare for and capture the opportunity
• High opportunity	Senior executive management attention needed, and management responsibility specified
• Medium opportunity	Manage the opportunity by specific monitoring or response procedures
• Low opportunity	Manage the opportunity by routine procedures. It is unlikely to need specific use of resources

3.5 Risk Evaluation

3.5.1 Risk Evaluation in the Risk Management Process

Through the previous step ("risk analysis") the risks have been identified, and based on the outcomes, we have to make decisions, about the risks that can be accepted or that need treatment. In this step, the "risk evaluation" process, we compare the level of the identified risks during the risk analysis process with the risk criteria defined when establishing the context (Sect. 3.2). So, risk evaluation will guide us to decide whether the risk is acceptable or unacceptable. Our understanding of the risk will support us to make decisions about future actions.

Decisions about future actions may include [3]:

- not to undertake or proceed with the event, activity, project or initiative
- actively treat the risk
- prioritising the actions needed if the risk is complex and treatment is required
- accepting the risk

Whether a risk is acceptable or unacceptable is a willingness to tolerate the risk; that is, a willingness to bear the risk after it has been treated to achieve the desired goals.

Risk attitude, appetite and tolerance are likely to vary over time, throughout or parts of the Organization as a whole.

A risk may be acceptable or tolerable in the following circumstances:

- No treatment is available
- Treatment costs are unaffordable (particularly in the case of lower ranked risks)
- The level of risk is low and does not warrant that using resources we can treat it
- The opportunities involved significantly outweigh the threats.

A risk is regarded as acceptable or tolerable if the decision has been made not to treat it (in accordance with the next step, "Treat risks", see Sect. 3.5.2), but remember that a risk considered as acceptable or tolerable does not imply that the risk is insignificant. **Risks that are considered acceptable or tolerable risks may still need to be monitored**.

3.5.2 Evaluation from Qualitative Analysis

Organizations have limited resources, so they must take advantage of opportunities or deal with their risks. So, Organizations need to define priorities. **Qualitative analysis is often used to set priorities or treatment based on the level of risk**.

As per the matrix presented in Sect. 3.4.5, the different levels of risk in a qualitative risk matrix can be used to define different actions required.

3.5.3 Tolerable Risk or Acceptable Risk

Tolerable risk or acceptable risk level of risk is the risk accepted in a given context based on the current values of society. The terms "acceptable risk" and "tolerable risk" are considered synonymous.[3]

When discussing tolerable risk or acceptable risk, the following questions may arise:

- What level of risk, if any, is acceptable in the workplace?
- What do we mean by "acceptable risk"?
- Can some risk be acceptable? If yes, acceptable to whom?

Remember: **Only Relative Safety Is Achievable**

All products and systems include hazards and some level of residual risk. However, the risk associated with those hazards should be reduced to a tolerable level. Safety is achieved by reducing risk to a tolerable level—"tolerable risk". Tolerable risk is determined by the search for an optimal balance between the ideal of absolute safety and the demands to be met by a product or system, and factors such as benefit to the user, suitability for purpose, cost-effectiveness, and conventions of the society concerned.

Developments, both in technology and in knowledge, can lead to economically feasible improvements to attain the minimum risk compatible with the use of a product or system.

[3]ISO/IEC GUIDE 51:2014.

3.5.4 The ALARP Principle

3.5.4.1 Introduction

ALARP is the acronym for "As Low As Reasonably Possible," a principle primarily used in the United Kingdom and Australia/New Zealand. In the U.K., the term "SFAIRP" is also commonly used, standing for "So Far As Is Reasonably Practicable." It has the same essential meaning as ALARP. The principles of ALARP are used in the United States in the area of nuclear radiation protection, where it is usually called "As Low As Reasonably Achievable" (ALARA). In the European Union, a similar classification is used (see [4] page nr. 40).

All business processes, which have a reasonably foreseeable risk of death or serious injury, should be assessed to ensure people are reasonably protected against significant risk of death, serious injury or other serious harm. When a new process is introduced, or an existing one is changed, a risk assessment should be done to determine acceptability. Each Organization can measure its performance using accident rates such as Fatal Accident Rate (FAR) and Equivalent Fatal Accident Rate (EFAR) per 100,000,000 exposure hours, or other appropriate measures [5].

The role of the risk assessment expert is to undertake the analysis, estimate the risk and its possible changes under various courses of action. There are several ways to make this decision. **Cost-Benefit Analysis** and **As Low As Reasonably Possible are two options.**

Next, the **ALARP principle** (Unacceptable risk, Tolerable risk and Negligible risk) is presented. The Australian/New Zealand Standard for Risk Management - AS/NZS 4360:2004 can be referred to.

The "As Low As Reasonably Practicable" or ALARP concept is illustrated in Fig. 3.7.

The diagram (carrot diagram) is shaped like a carrot, wider at the top and pointed at the bottom, and is divided into three regions: generally unacceptable, tolerable, and acceptable. It allows industries to visually represent risks, as well as recognize risks that should be further reduced.

As mentioned, the ALARP principle recognizes that there are three broad categories of risks:

1. **Unacceptable risk**: An upper band where adverse risks are intolerable whatever benefits the activity may bring, and risk reduction measures are essential whatever their cost. The risk level is so high that we are not prepared to tolerate it. The losses far outweigh any possible benefits in the situation.
2. **Tolerable risk**: A middle band (or 'grey' area) where costs and benefits, are taken into account and opportunities balanced against potential adverse consequences. Risk is tolerable in view of the benefits obtained by accepting it. The cost in inconvenience or in money is balanced against the scale of risk and a compromise is accepted. This would apply to travelling in a car, we accept that accidents happen, but we do our best to minimize our chances of disaster. Does it apply to bungee jumping?

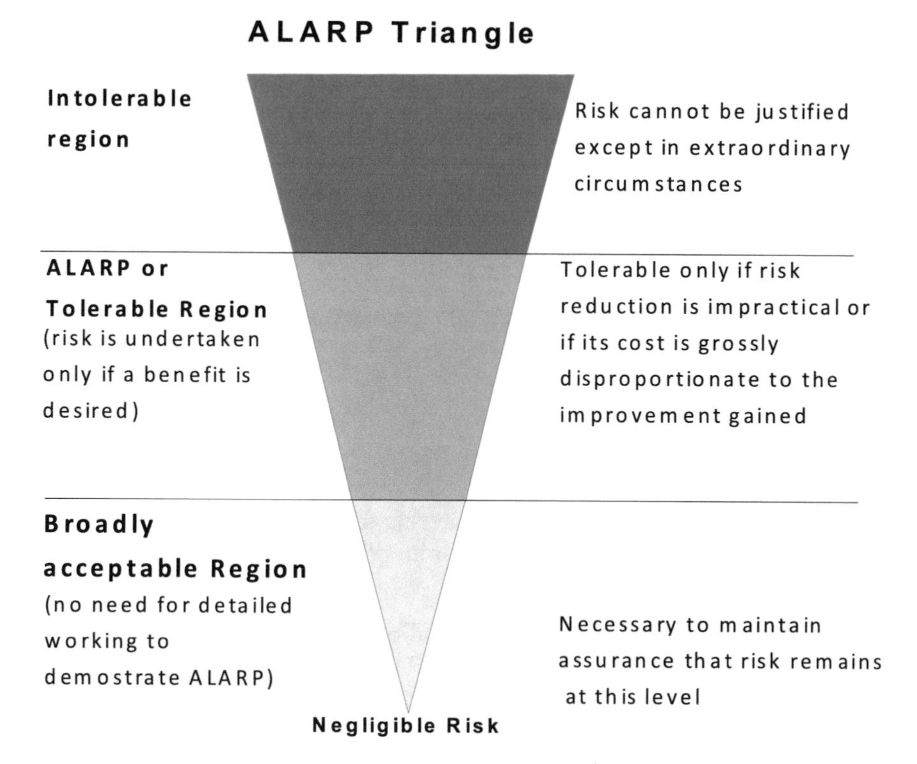

Fig. 3.7 The ALARP principle

3. **Negligible risk**: A lower band where positive or negative risks are negligible, or so small that no risk treatment measures are needed. Risks are broadly accepted by most people as they go about their everyday lives, these would include the risk of being struck by lightning or of having brake failure in a car.

Let us examine the following cases:

- risks are close to the intolerable level: the expectation is that risk will be reduced unless the cost of reducing the risk is unacceptably disproportionate to the benefits gained.
- risks are close to the negligible level: action may only be taken to reduce risk where benefits exceed the costs of reduction.

The two aspects of cost and benefit need to be balanced. The question "*Can something be done*?" shall be balanced against the "*Is it worth doing something in the circumstances*?".

ALARP principles apply until it is demonstrated that further risk reduction in infeasible or that the cost of reducing the risk further is disproportionate to the benefit.

Here an example on the application of the ALARP Principle.[4]

A 1,000,000 USD cost per "life saved" target is used in a particular industry. A maximum tolerable risk target of 10^{-4} pa (per annum) has been set for a particular hazard which is likely to cause 2 fatalities. The proposed system has been assessed and a predicted risk of 8×10^{-5} pa obtained. Given that the negligible risk is taken as 10^{-6}pa then the application of ALARP is required. For a cost of 3,000 USD, additional instrumentation and redundancy will reduce the risk to just above the negligible region (2×10^{-6} pa). The plant life is 30 years.

Solution:

$1.000.000 \times (8 \times 10^{-5} - 2 \times 10^{-6}) \times 2 \times 30 = 4680$ USD > 3.000 USD. Therefore the proposal should be adopted.

3.5.4.2 Upper and Lower Bounds for Risk

One of the problematic aspects of implementing the ALARP principle is to define the three-level of risks. Here the approach from HSE (Health and Safety Executive, in the UK).[5]

HSE believes that an individual risk of death of one in a million per annum for both workers and the public correspond to a very low level of risk and should be used as a guideline for the boundary between the broadly acceptable and tolerable regions.

Nevertheless, in our document on the tolerability of risks in nuclear power stations, we suggested that an individual risk of death of one in a thousand per annum should on its own represent the dividing line between what could be just tolerable for any substantial category of workers for any large part of a working life, and what is unacceptable for any but fairly exceptional groups. For members of the public who have a risk imposed on them 'in the wider interest of society,' this limit is judged to be an order of magnitude lower – at 1 in 10,000 per annum.

Essentially this principle guides the design engineer and the safety specialist into setting tolerable risk targets for a hazardous situation. This is the first step in setting up a standard of performance for any safety system.

The upper and lower bounds for risk in each Organization relate to an individual's exposure, i.e. how likely is one individual to die in one year.

This is equivalent to the total number of deaths per annum divided by the number of people in the exposed population.

An example of upper and lower bounds for risk in an Organization is given in the Table 3.13 [5].

Risk above the upper bound is **Intolerable** and must be dealt with immediately, including temporarily creating an activity until improvements can be made.

Risk within the upper and lower bounds is **Tolerable** but should be "Reduced At Reasonable Cost" (ALARP). These risks should be subject to cost/benefit

[4]http://www.gt-engineering.it/en/Insights/the-alarp-principle.

[5]http://www.gt-engineering.it/en/Insights/the-alarp-principle.

Table 3.13 Upper and lower bounds for risk in an Organization [5]

	Deaths per annum		Equivalent deaths per annum	
	Upper bound	Lower bound	Upper bound	Lower bound
Rail personnel	1 in 1,000	1 in 1,000,000	1 in 400	1 in 400,000
Passengers and public (excludes illegal acts)	1 in 10,000	1 in 1,000,000	1 in 4,000	1 in 400,000

calculations to determine the value of undertaking risk mitigation steps. Fatalities and serious injuries are costed using the statistical value of avoided deaths and injuries criteria based on "willingness to pay" research.

Risk less than the lower bound is considered **Acceptable**.

This is illustrated with reference to the ALARP triangle in the Table 3.14 on the following Section (Sect. 3.5.4.3).

3.5.4.3 ALARP Criteria

DPA Death per annum
EDPA:Equivalent Deaths per annum (Sect. 4.1.10).
EFAR:Equivalent Fatal Accident Rate (Sect. 4.1.10).
FAR: Fatality Accident Rate (Sect. 4.1.10).

3.6 Risk Assessment

The previous three steps described—Identify the risk, Analyse the risk and Evaluate the risk—form the **Risk Assessment phase** of the risk management process (see Figs. 3.1 and 3.8).

3.7 Risk Treatment

3.7.1 Introduction

Risk treatment is a step in the risk management process which follows the phase of risk assessment. In the risk assessment, all the risks have been identified, and risks that are not acceptable must be selected. The main task in the risk treatment step is to select one or more options for treating each unacceptable risk, i.e. decide how to

Table 3.14 ALARP criteria (as per [1])

ALARP triangle		Accident rate thresholds							
		Rail personnel				Public and passengers			
		DPA	EDPA	FAR	EFAR	DPA	EDPA	FAR	EFAR
Intolerble region	Risk cannot be justified except in extraordinary circumstances	1 in 1,000	1 in 400	50	125	1 in 10,000	1 in 4,000	20	50
ALARP or Tolerable Region (risk is undertaken only if a benefit is desired)	Tolerable only if risk reduction is impractical or if its cost is grossly disproportionate to the improvement gained. Tolerable if cost of reduction would exceed the improvement gained	1 in 100,000	1 in 40,000	0.50	1.25	1 in 100,000	1 in 40,000	2	5
Broadly acceptable region (no need for detailed working to demonstrate ALARP)	Necessary to maintain assurance that risk remains at this level	1 in 1,000,000	1 in 400,000	0.05	0.125	1 in 1,000,000	1 in 400,000	0.20	0.30

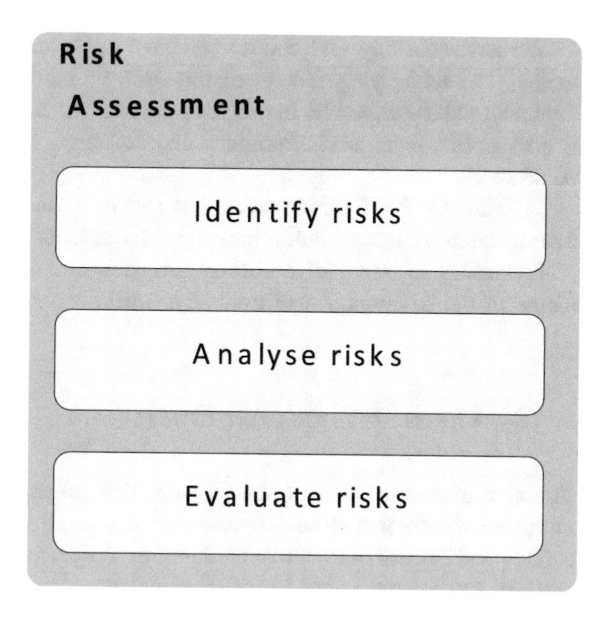

Fig. 3.8 Risk assessment

mitigate all these risks, evaluating those options, preparing treatment plans and implementing them.

The previous steps of the risk management process will help us to get a comprehensive understanding of the risks concerned. During the risk treatment process, it is important to **identify the causes of the risks so that these risks are treated and not just the symptoms**. We have to understand how risks arise, understand not only the immediate causes of an event.

In general, there are five types of risk treatment:

1. **Avoidance**: You can choose not to take on the risk by avoiding the actions that cause the risk. For example, if you feel that driving in the night is too dangerous, you can avoid the risk by driving only during the day.
2. **Reduction**: You can take mitigation actions that reduce the risk. For example, wearing a seat belt when driving.
3. **Transfer**: You can transfer all or part of the risk to a third party. The two main types of transfer are insurance and outsourcing. For example, a company may choose to transfer project risks by outsourcing the project.
4. **Acceptance**: Risk acceptance, also known as risk retention, is choosing to deal with a risk. In general, it is impossible to make business without choosing to take on risk. For example, an investor might accept the risk of a company going bankrupt when he buys its bonds.
5. **Sharing**: Risk sharing is the distribution of risk to multiple Organizations or individuals. This is done for a variety of reasons, including insurance products and self-insurance strategies.

We have to study and choose the most appropriate combination of risk treatments. The underlying factors of the risks will influence whether the proposed treatment will be effective. In this regard, costs and benefits, effectiveness and other factors as like legal, social, political and economic considerations may need to be considered.

Individual risks should be treated as part of an overall treatment strategy, rather than isolated, to ensure that critical dependencies and linkages are considered.

Figure 3.9 in Sect. 3.7.2 outlines the risk treatment process and the iterative nature of the development of treatment action plans.

3.7.2 The Risk Treatment Process

The risk treatment process shall ensure that effective strategies are in place to minimise the frequency and severity of the identified risk. We need to develop actions and implement treatments that aim to control the risk.

Once the risk assessment phase is complete, identify the options for treatment if there are any; otherwise, tolerate the risk. Where treatment options are available and appropriate, document such treatment options as part of the risk treatment plan.

Treatment options not applied to the source or root cause of a risk are likely to be ineffective and promote a false belief within the Organization that the risk is controlled.

Figure 3.9 outlines the risk treatment process, and in the next sections, the steps of the process are described in detail (as per [3]).

3.7.2.1 Decide if Specific Treatment Is Necessary

Decide if specific treatment is necessary or whether the risk can be adequately treated in the course of standard management procedures and activities; that is, embed the treatment into day-to-day practices or processes. When deciding what treatments could be implemented, it is useful to consider ways in which standard practices already serve as controls, or ways in which those standard practices could be changed to better control the risk.

3.7.2.2 Develop a Risk Treatment Plan

Work Out What Kind of Treatment Is Desirable

Work out **what kind of treatment** is desirable for this risk—determine what the goal is in treating this particular risk; is it to avoid it completely, reduce the likelihood or consequence, transfer the risk (to someone else such as an insurer or contractor) or accept the level of risk based on existing information? The type of

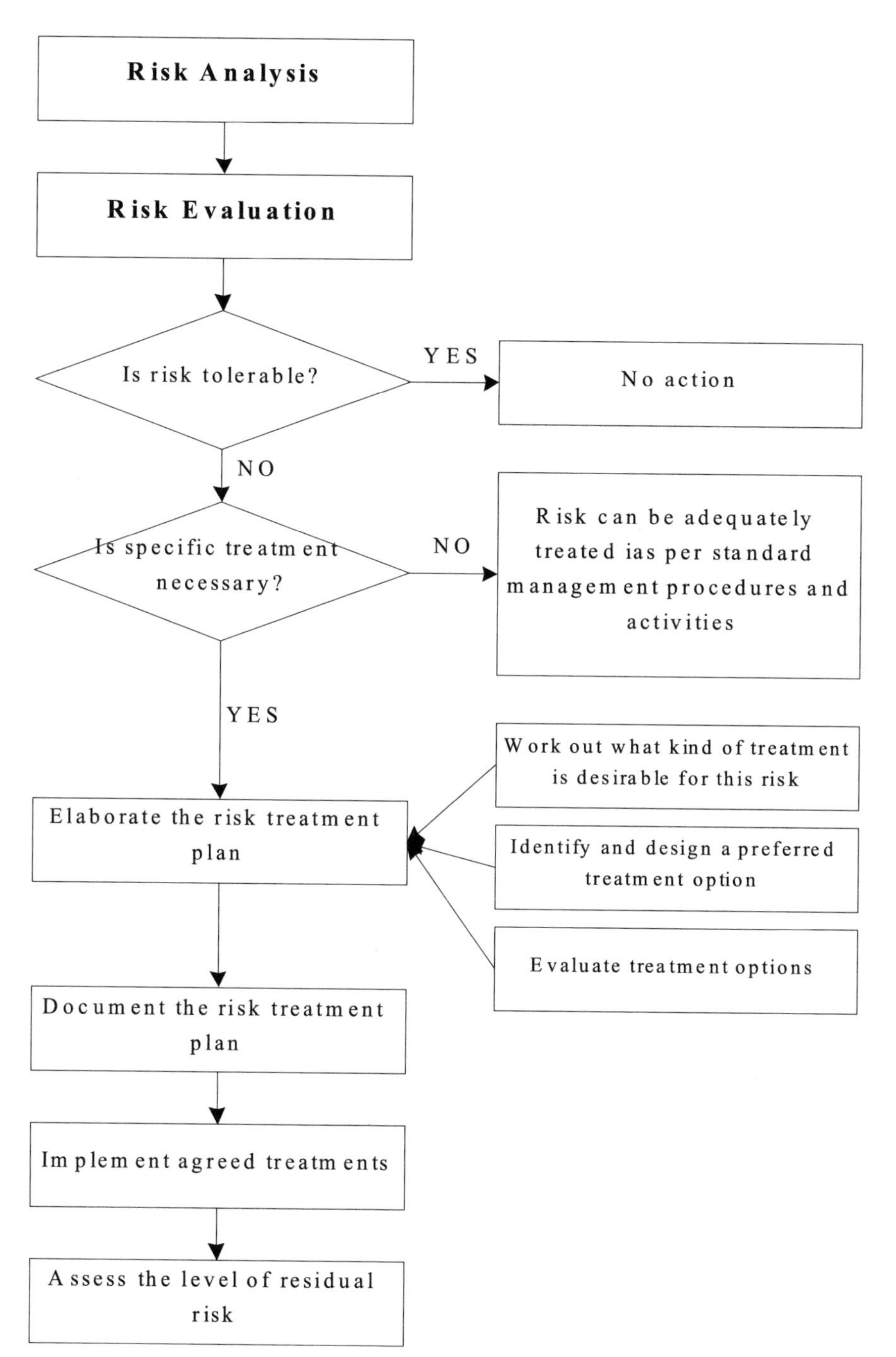

Fig. 3.9 The risk treatment process

risk treatment chosen will often depend on the nature of the risk and the tolerance for that risk.

Identify and Design Preferred Treatment Options

Identify and design a preferred treatment option once the goal of treatment is known.

If the goal is to **reduce the likelihood or possibility of the risk**, then you may need to adjust what is happening or might be planned: successfully altering the approach will depend on identifying the causes of the threat and the causal links between the threat and its impact—both of which should have been identified in the risk assessment phase.

If it is not possible to change the approach of the project or activity, then some further action may be taken to minimize the occurrence of the event or to reduce the likelihood of a threat.

Understanding the nature of the risk event and how it occurs, will make it easier to identify any possible measures that would reduce the risk.

If the goal is to **reduce the consequence or impact of the risk**, then contingency plans might be required to respond to a threatening event if it occurs. This planning may be undertaken in combination with other controls—that is, even if steps have been taken to minimise the likelihood of the risk, it may still be worthwhile to have a plan in place to reduce the consequences if the event actually occurs.

If the goal is to **share the risk**, then involving another party, such as an insurer or contractor, may help. Risk can be shared contractually, by mutual agreement, and in a variety of ways that meet all parties' needs. Any such arrangement should be formally documented—whether through a contract or agreement or by letter.

Sharing the risk does not remove our obligations and does not avoid us suffering consequential damage if something unexpected happens or something goes wrong.

If the risk is so significant that the objective is to reduce or completely avoid it, then the options are limited to significantly changing the project, selecting alternative approaches or procedures to make the risk negligible, or leaving the activity or project partner or programme. It is not often possible to eliminate a risk completely and balance is an important part of the risk assessment exercise.

Sometimes a decision is made to accept or tolerate the risk because of the low likelihood of minor implications of the risk event, or because the cost of controlling the risk efficiently is unjustifiably high or the opportunity outweighs the risk. The Organization recognizes that taking calculated risks is both acceptable and appropriate in achieving its strategic goals. However, the decision to accept risk should be carefully documented in these cases, so that a record will be available for future reference (or evidence) if the risk does occur. Contingency planning should also be considered for dealing with and reducing the consequences, should they arise.

As mentioned in Sect. 3.7.1, a combination of treatment options shall be chosen from the defined range of options. The options selected need to be compatible with the overall objectives of the Organization and with the risk evaluation criteria.

When selecting options, the following issues should be considered:

- Some benefits arising from the treatment may be more important than others.
- Non-quantifiable benefits and costs may at times be considered more important than quantifiable ones. In this case, decisions should not be based exclusively on quantitative analysis.
- Direct and indirect benefits and costs associated with risk treatments may occur over different time periods, and this should be considered in any quantitative and qualitative analysis.
- Various levels of uncertainty can affect estimates of direct and indirect benefits and costs and may follow different probability distribution curves.
- Social expectations, as well as legal obligations, may impose specific risk treatment actions.
- Often there is a particular aversion to events which represent "human dread".
- For events that conflict with organizational values and that can harm the Organization's reputation and image, there is an analogous type of "corporate dread".

Evaluate Treatment Options

Evaluate the treatment options and determine their effectiveness with respect to risk tolerance. Do the controls selected appear to have the desired treatment effect (that is, will the controls stop or reduce what they are intended to stop or reduce)?

Will the controls trigger any other risks? For example, a sprinkler system designed to address the fire risk may cause water damage, creating a different risk that requires consideration or management.

Are the controls beneficial or cost-efficient? Does the cost of implementing the control outweigh the costs that would result from the event that occurs without the control in place? Overall, is the cost of implementing the control reasonable for this risk?

The process of treating a risk, determining if residual risk levels are tolerable, and evaluating the effectiveness of that treatment are all case - by - case assessments that rely on a clear understanding of the risk and a focus on the ultimate objective of the activity being assessed.

3.7.3 Document the Risk Treatment Plan

As presented, once the treatment options have been identified, a risk treatment plan should be prepared. It can be easily generated through the Organization risk register once a risk is recorded. Treatment plans should identify responsibilities for action, time frames for implementation, budget requirements or resource implications, performance measures and review process where appropriate. The review process should monitor the progress of treatments against critical implementation milestones.

3.7.3.1 Implement Agreed Treatments

Once any options requiring authorization have been approved for resourcing, financing or other actions treatments should be initiated by those recognized as having the responsibility to do so. Finally, the person entrusted with primary risk responsibility is responsible for the risk treatment.

3.7.3.2 Assess Residual Risks

(see https://advisera.com/27001academy/knowledgebase/why-is-residual-risk-so-important/)

Residual risk is "the risks remaining after risk treatment".

Once the risk has been treated, the level of residual risk needs to be assessed. Even when a risk has been treated and the controls are in place the risk may not be completely eliminated. The level of residual risk refers to the likelihood and consequence of the risk occurring after the risk has been treated. The residual risk rating is generally lower than the initial risk rating; otherwise, the controls were not effective.

The Organization needs to know precisely whether the planned treatment is enough or not. Residual risks are usually assessed in the same way as we perform the initial risk assessment—we use the same methodology, the same assessment scales, etc. What is different is that we have to consider the influence of controls (and other mitigation methods), so the likelihood of an incident is usually reduced, and sometimes the impact is even lower. All residual risks should be documented and should be monitored and reviewed.

As we discussed, the purpose of identifying residual risks is to find out whether the planned treatment is sufficient, but how we know what it is sufficient? This is where the **concept of an acceptable level of risks** comes into the picture—it is nothing else but deciding how much "**risk appetite**" an Organization has, or in other words whether the management thinks it is fine for an Organization to operate in a high-risk environment where it is much more likely that something will happen, or the management wants a higher level of security involving a lower level of risk.

Every Organization has to decide what is appropriate for its circumstances (and for its budget).

Residual risk management: Once you find out what residual risks are, what do you do with them? Basically, you have these three options:

- If the level of risks is below the acceptable level of risk, then you do nothing—the management needs to accept those risks formally.
- If the level of risks is above the acceptable level of risk, then you need to find out some new (and better) ways to mitigate those risks—that also means you'll need to reassess the residual risks.
- If the level of risks is above the acceptable level of risk, and the costs of decreasing such risks would be higher than the impact itself than you need to propose to the management to accept these high risks.

On that way, the top management of the Organization is involved in reaching the most important decisions, and that nothing is ignored: so, the point is that top management needs to know which risks the Organization will face even after various mitigation methods have been applied.

3.7.3.3 The Risk Treatment Plan or Strategy

Too much treatment is as undesirable as also too little if the resources of an Organization and the management attention is diverted from more business-critical activities.

After selecting the proper risk treatment options for our risks, risk treatment plans or strategies and action plans should be elaborated. Treatment actions for different risks need to be combined and compared to identify and resolve conflicts and eliminate redundancies.

Treatment plans should:

- Identify responsibilities, schedules (incl. critical milestones), the expected outcomes, the required budget for risk treatment, performance indicators and the review process.
- Include procedures for assessing and monitoring treatment effectiveness against treatment objectives and organizational objectives, and monitoring treatment plan progress against critical milestones.
- Document how the chosen options will be implemented.

The risk treatment plan needs to be part of an effective management system. Communication is an essential part of the risk treatment plan.

3.7.4 Contingency Planning

If an adverse risk occurs, contingency planning is the act of preparing a plan or a series of activities. Having a contingency plan in place forces the Organization (project teams, etc.) to consider a course of action in advance if a risk event occurs.

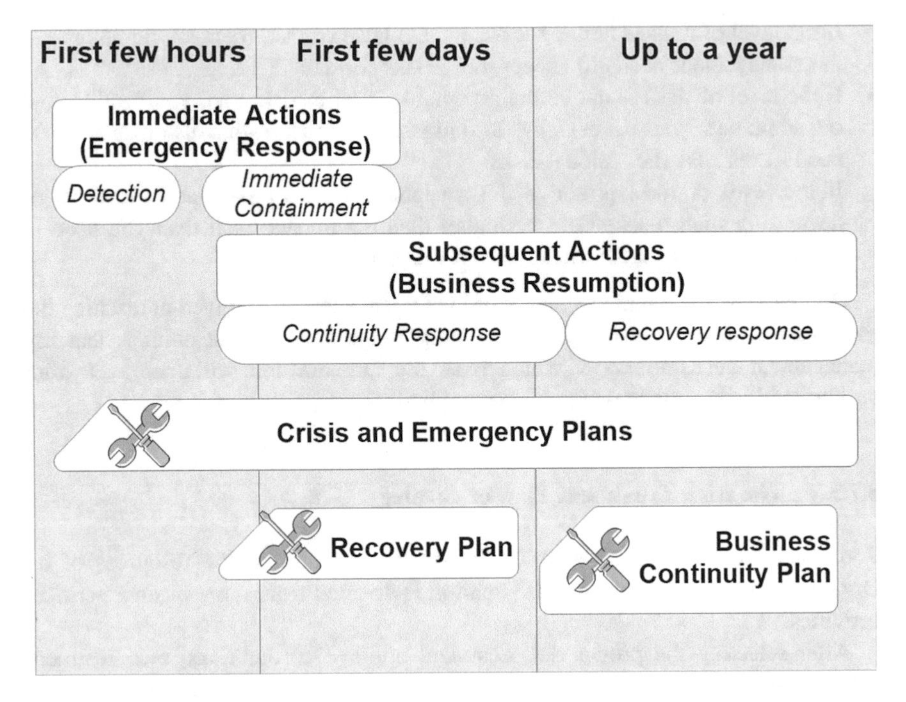

Fig. 3.10 Managing critical incidents (based on [2])

Contingency can also be reflected as a line item in the budget of the Organization, to cover unexpected expenses. The amount to budget for contingency may be limited to just the level of the probability of risks.

This is usually determined by estimating the cost when a risk occurs and multiplying it by the probability. For example, if a risk is estimated to result in an additional cost of $ 500,000 and the probability is 80%, the amount to be included in the budget of this one item is $ 400,000.

In case of an event, the management of the Organization must respond quickly in order to mitigate the impact of the event on the goals of the Organization. These impacts can normally be managed as part of the normal processes of management. However, if the scale of the event exceeds the Organization's capabilities, a systematic approach to critical incident management is required (Fig. 3.10).

Business Continuity Management (BCM) is a key element of critical incident management. BCM is a framework for identifying the risk of an Organization being exposed to internal and external threats. BCM's goal is to give the Organization the ability to respond effectively to threats and protect the Organization's business interests. BCM includes recovery from disasters, business recovery, crisis management, management of incidents, emergency management and contingency planning.

3.7.5 *Sharing/Transferring Risks*

Risk sharing is a risk management method in which the cost of the consequences of a risk is distributed among several participants, i.e. another Organization(s) who is bearing or sharing some or all of the risk, usually via a contract. The risk has not gone away, but another party shares the responsibility for managing that risk. Subcontracting, outsourcing and insurance are the most common forms of risk-sharing.

3.7.5.1 Contracting

Contracting transfers to the vendor specific risks, depending on the work required by the contract.

Contracting principles are to allocate risk through a contract document to the party best placed to manage it. Inappropriate transfer of risk may result in a change in the nature of the risk or in emerging of new risks.

3.7.5.2 Insurance

We all know how insurance works. Car insurance is a good example of risk transfer. We buy car insurance so the insurance company will pay the cost of repairing the damage to the car in the event of an accident. So, in order to cover the cost of a damage to our car, we do not need adequate own financial resources.

Similarly, in most cases, our Organizations do not have sufficient internal financial resources to cover the cost of major losses. Insurance is a way to arrange to fund by paying a premium for specific types of losses. For example, insurance cover may be available for property damage and consequential costs (such as income loss or increased operating costs) or liability for other party's financial consequences due to failure to discharge a legal obligation.

Insurers consider for the calculation of the premium for a particular risk the likelihood and consequence and take account of existing controls. In some cases, insurers may request additional treatment to reduce the risk, or they may agree to charge less if additional inspections are carried out or if the insured party agrees to share the risk, i.e. by paying the first part of any loss to an agreed amount or excess (deductible).

Not all major risks are insurable. For example, the reputation of the Organization cannot generally be insured, but the loss of trust by its Stakeholders could be fatal for some Organizations.

3.8 Monitor and Review

3.8.1 Introduction

Having identified, recorded, analysed risks and initiated the agreed treatments, an appropriate monitoring and reporting regime must be established for effective treatment and risk control.

Risk management processes should be incorporated into the organizational processes. Risk management is a dynamic process, and changes as the Organization changes. Monitor focuses on changes in the risk source and context, tolerance for certain risks, and appropriateness of controls. Risks, therefore, need to be reviewed and regularly reported. The frequency of the review will depend on the risk rating, strength of control, and the ability to effectively treat the risk.

Monitoring and review are an essential and integral part of risk management: it is possible to evaluate the effectiveness and appropriateness of strategies and management systems for implementing risk treatments and the risk management plan as a whole by monitoring and reviewing the risks.

It is also necessary to identify emerging risks as soon as possible and to monitor them as well as the known risks.

Periodic reviews of risk and treatment strategies should be linked to development of business plans and strategic plans. When organizational changes are planned, or external changes are detected there may be changes in the organizational context (the objectives, the internal or external environment, risk criteria etc.), changes in the risks and levels of risk as also the effectiveness of risk treatments.

Priority should be given to monitoring:

a. High risks.
b. Failure of treatment strategies, especially where this would result in high, or frequent, consequences.
c. Risk-related activities.
d. Risk tolerance criteria especially where this results in high levels of residual risk.

How frequently a review process and reporting cycle occurs will depend on the risk appetite and level of risk tolerance.

3.8.2 Types of Monitoring and Review

There are three types of monitoring and review:

- Frequent monitoring through regularly measuring or checking of specific parameters
- Line management review
- Internal and external auditing.

3.8.2.1 Continuous Monitoring

As discussed, when risk treatments are agreed, an appropriate monitoring system should be established to provide ongoing evidence that treatments are effective.

The risk register (see Sect. 4.2.2) is a basis for monitoring. As the risk register keeps record of existing controls and treatments, it provides a good basis when setting up a monitoring regime. Issues discussed in Sect. 3.8 should be considered. It is essential to keep the risk register always updated.

3.8.2.2 Line Management Review

It is the responsibility of line managers to regularly review processes and activities in all areas of their responsibility. The aim of regular review is to identify when new risks arise, monitoring existing risks to ensure that current treatment strategies remain effective and appropriate.

If any issues are identified, the risk management systems should be examined for any indication of general weakness. If there is such an indication, it may require a more detailed review.

3.8.2.3 Third-Party Audit

Third-party audits are an independent measure.

- **Internal audit**: The internal audit program of the Company has as its goal the review of the system, policy and process assurance. The auditors apply a risk - based approach to the audit program and help to bringing a measure of independence and external perspective to the Organization's risk management framework.ss
- **External audit**: external audit usually covers the systems and processes of financial, governance, contracting, IT and risk management. Other audits may occur from time to time and are subject to contracts, the legal framework for the country, etc.

Audit findings will typically indicate systemic weakness. Response to the audit should concentrate on remedying the system and raising the causes and not simply the symptoms.

3.8.3 Risk Management Performance

Key Performance Indicators (KPIs) are quantitative measures to monitor an activity's level of performance or to the Organization as a whole. They must be measurable and tailored to individual units of the Organization. KPIs should help in continuous improvement and the employees should be responsible for managing this continuous improvement process.

KPIs should primarily focus on:

- The highest risks.
- The most critical treatments or other processes.
- Treatments or processes with the greatest potential for efficiency improvements.

Some examples of useful risk management performance indicators are [2]:

- A decline in the total cost of risk.
- Progress towards a specific organizational objective.
- The extent to which recommendations for risk treatment are implemented.

3.8.4 Post-event Analysis

> Mistakes have the power to turn you into something better than you were before. (Anonymous)
>
> When you make a mistake, there are only three things you should ever do about it: admit it, learn from it, and don't repeat it. (Paul Bear Bryant)
>
> You can't make the same mistake twice. The second time you make it, it's no longer a mistake. It's a choice. (Anonymous)

Events, as like failures, incidents, accidents will always occur, no matter how well an Organization manages risks. Risk management will not stop the occurrence of events/issues/incidents, but Organizations with a well-developed risk management framework will learn from their mistakes. Such events are likely to happen again without a well-structured risk management program in Organizations.

Events but also successes can contribute to improving the risk management process, by monitoring and reviewing risks and treatments. Organizations should adopt a systematic process to review the causes of successes, failures, and near-misses and to learn from their mistakes.

The post-event analysis itself is going to ask some fundamental questions[6]:

- What happened?
- Why did it happen? Did we previously identify and analyse the risks involved? Did we identify the actual causes in risk identification?
- Did we rate and assess risks and controls correctly?
- Could we have done anything to prevent that event?

[6]https://paladinrisk.com.au/risk-tip-10-post-event-analysis/.

- Did the controls operate as intended?
- Were the treatment plans effective? If not, where could improvements be made?
- Were our monitoring and review processes effective?
- Is there anything we can do to prevent the incident occurring again in the future?
- If the event does occur again in the future, are there any strategies we can put in place to minimise the impacts?
- What do we need to do to ensure that failure events are not repeated but that successes are?
- How could our risk management process, in general, be improved?

Think about: **Today's incident is yesterday's and tomorrow's risk.**

3.9 Communicate and Consult

3.9.1 Introduction

What is **communication**? There are several definitions of risk communication. We define here risk communication as the "*exchange of information and opinions, and establishment of effective dialogue, among those responsible for assessing, minimizing, and regulating risks and those who may be affected by the outcomes of those risks*".[7]

Risk management can be improved during every stage of the risk management process through effective communication and consultation with internal and external Stakeholders. The Stakeholders must be identified early, as discussed in Sect. 3.2.2.

All parties must understand each other's perspectives and take an active part in the decision-making process, where appropriate. It is important to ensure that those responsible for risk management, but also interested stakeholders, understand the basis on which decisions are made and why specific treatment options are chosen.

Risk communication will not solve all of an Organization's problems or resolve all conflicts. Inappropriate communication of risk can lead to a collapse of trust and therefore, poor risk management.

Consultation is the process of "*seeking and giving of advice, information, and/ or opinion, usually involving a consideration*".[8] It is a process of informed communication between the Organization and its Stakeholders on an issue before a decision or a direction on a particular issue is taken [2].

Communication and consultation should be considered at each step of the risk management process. As discussed in Sect. 3.2 (Establish the context), it is essential to identify our Stakeholders and consider their needs. A communications plan to be developed after the identification of the Stakeholders (Sect. 3.2.2). This

[7]http://www.businessdictionary.com/definition/risk-communication.html.

[8]http://www.businessdictionary.com/definition/consultation.html.

plan should specify the purpose or goal for the communication, who is to be consulted and by whom, when it will take place, how the process will occur, and how it will be evaluated.

Good communication can contribute to the development of a "culture" where it recognizes and estimates the positive and negative aspects of risk. Communication can help an Organization develop its risk attitude.

An Organization might find it inappropriate for commercial or security reasons to communicate with stakeholders. The communication plan should document the decision not to involve Stakeholders in these situations, but its requirements and perspectives should be considered.

3.9.2 Risk Perceptions by Various Stakeholders

Risk perceptions may vary due to differing values, experiences, beliefs, hypotheses, needs and concerns of the Organization's employees and various Stakeholders. Since Stakeholders can have a significant impact on risk management activities (especially Stakeholders with high influence/high interest as per Stakeholder analysis (Sect. 3.2.2), it is important to identify and record their perceptions of risk and to understand and address the reasons for their perceptions of risk.

Since perceptions may vary, communicating the level of risk effectively is essential. For example, if people live near a hazardous site, then the understanding of risk can be enhanced by involving the resident community in risk management aspects. Effective communication will, therefore, help develop an association with an Organization's external Stakeholders and establish trust-based relationships. Similarly, trust is also essential within the Organization.

Involvement of the Stakeholders brings a greater diversity of approaches and perceptions to risks and views about objectives. Where there is high uncertainty, the beliefs and values of people are fundamental.

The experience and expertise of the various internal and external Stakeholders will often improve understanding of the risk. Perception diversity can improve risk assessment and avoid "group thinking," resulting in irrational or non-optimal decisions. Senior management, for example, may have a different view of the consequences of risk compared to other levels of management. Workers may see risks that others oversee and may have a better understanding of risk events than their managers. Taking ownership of risk treatment decisions can ensure the acceptance of recommended treatments.

Risk communication can, therefore, be an important parameter in the process of risk treatment.

3.9.3 Communication and Consultation in the Risk Management Process

Communication and consultation are important considerations at all stages of a risk management process. The process should set up an Organization dialogue with its Stakeholders, a process of consultation rather than a one-way flow of information from decision-makers to other Stakeholders. As mentioned, all those responsible for implementing risk management, as well as those with a vested interest, need to develop a common understanding of how decisions are made (the decision mechanism) and why specific actions are needed.

A communication plan needs to be drawn up at the earliest stage of the process for both internal and external Stakeholders to address risk issues and the process for managing them.

A consultative team approach is useful in helping to properly define the context, helping to ensure that risks are effectively identified, bringing together different areas of expertise in risk analysis, ensuring that the risk management process is appropriately considered in different views.

If parties understand the perspectives of each other and are actively involved in the decision-making process where appropriate, risk management can be improved.

Thus, proper communication and consultation are aimed at:

- helping internal and external Stakeholders to understand the risk and risk management process
- ensuring Stakeholder views are considered and
- ensuring that all participants are aware of their roles and responsibilities and take ownership of risk management decisions and responsibilities.

3.9.4 The Communication and Consultation Plan

The first step in ensuring successful communication is the establishment of a Project Communication Plan.

A Communication Plan is a document that provides information to Stakeholders to help the Organization achieve specific goals. It describes how information will be disseminated to all relevant Stakeholders and received from them. An effective Communication Plan formally defines what information should be provided to specific segments of the audience, when that information should be provided, who is authorized to communicate confidential or sensitive information and how information (email, websites, printed reports and/or presentations) should be disseminated. Finally, the plan should define what communication channels will be used by Stakeholders to request feedback and how to document and archive communication.

Effective communication between Stakeholders is a critical risk management success factor in ensuring understanding of the context, identifying and assessing risks, and planning and owning appropriate responses. Like all communications plans, bidirectional communication is essential, so the plan must outline processes for handling feedback as well as information on the messages to be sent.

References

1. Australian/New Zealand Standard (2004) AS/NZS 4360 SET Risk Management
2. Standards Australia/ Standards New Zealand (2005) Risk Management Guidelines—Companion to AS/NZS 4360:2004
3. Organization of Adelaide, Australia, Risk Management Handbook
4. ERA (European Railway Agency) (2009) Collection of examples of risk assessments and of some possible tools supporting the CSM Regulation
5. New Zealand (2007) National Rail System Standard/4—Risk Management

Chapter 4
Risk Assessment Techniques

4.1 Techniques for Risk Assessment

4.1.1 An Overview

Various tools and methods are used for the assessment of risk. Some of these tools and methods are here presented.

As presented in Sect. 3.1.2, risk assessment includes (Fig. 4.1):

- Risk identification
- Risk Analysis
- Risk Evaluation

4.1.2 Risk Assessment Techniques as Per ISO/IEC 31010:2009

"*ISO/IEC 31010:2009 Risk management—Risk assessment techniques*" is a standard dedicated to risk assessment techniques. It is a supporting standard for "*ISO 31000 Risk management—Principles and guidelines*" and guides how to select and apply systematic techniques for risk assessment. It includes 31 different techniques, although some techniques converge. It is not critical that managers know all of them but knowing more about these techniques will help managers better align the risk assessment process with the risk assessment objectives.

The risk assessment techniques can be classified as following (Fig. 4.2)

- risk identification
- risk analysis

K. Tzanakakis, *Managing Risks in the Railway System*, Springer Tracts on Transportation and Traffic 18, https://doi.org/10.1007/978-3-030-66266-0_4

Fig. 4.1 Risk assessment

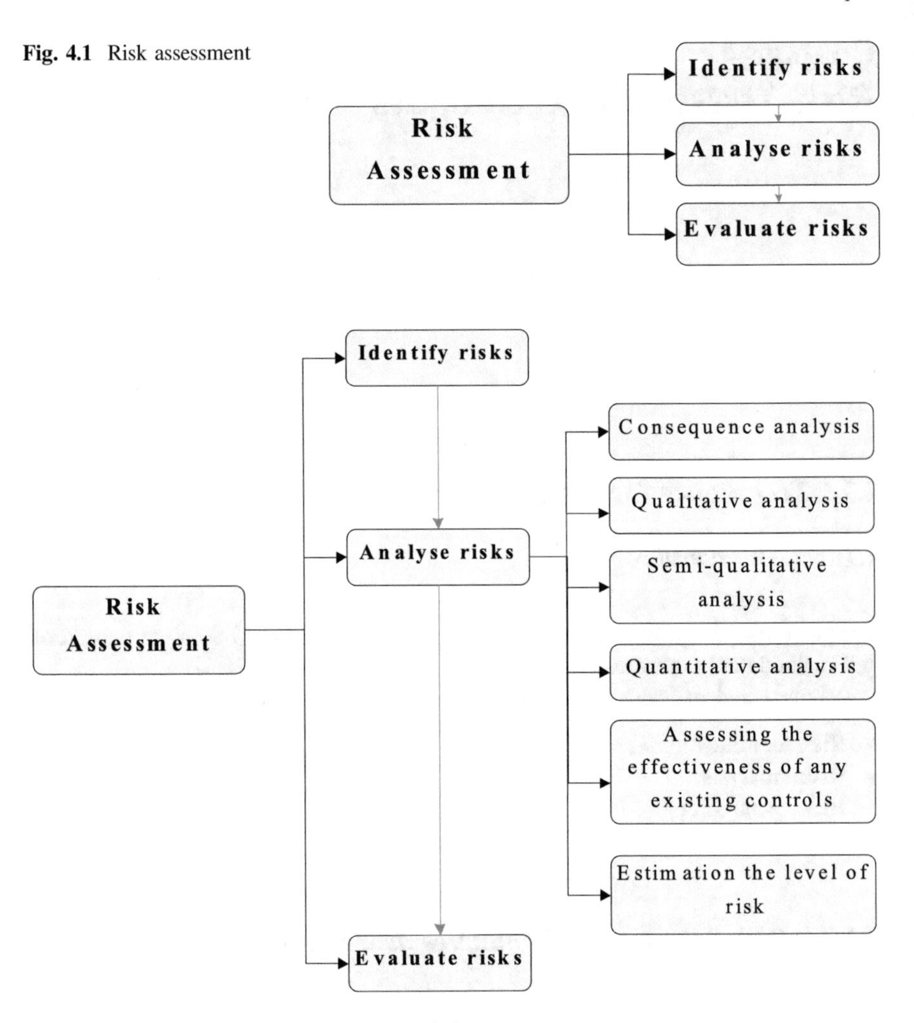

Fig. 4.2 Classification of risk assessment techniques

- consequence analysis
- qualitative,
- semi-quantitative or
- quantitative probability estimation
- assessing the effectiveness of any existing controls
- estimation the level of risk

- risk evaluation

Next Fig. 4.3 presents the various types of risk assessment techniques as per ISO/IEC 31010:2009.

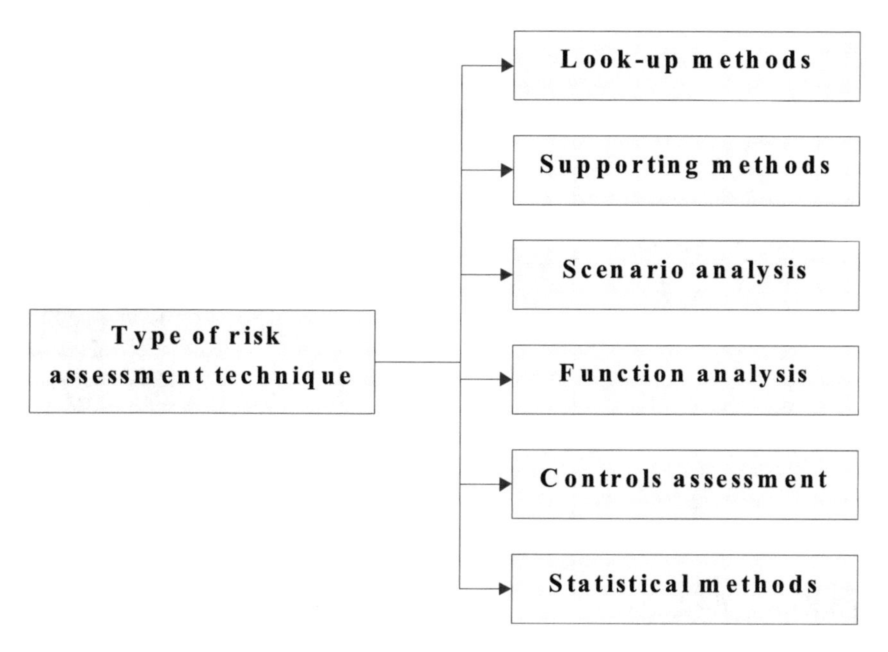

Fig. 4.3 Types of risk assessment techniques

The techniques suitable for the various risk assessment types are presented in Fig. 4.4. Techniques providing quantitative outputs are marked with a bold frame.

Table 4.1 presents the applicability of tools used for risk assessment as per ISO/IEC 31010:2009 (Annex A, Table A). The techniques described in the next sections are indicated in Table 4.1 with a reference to the related Section. In Annex A of ISO, all techniques are described in detail. Column 7 of Table 4.1 is referring to the related Section of the ISO.

4.1.3 Look-Up Methods

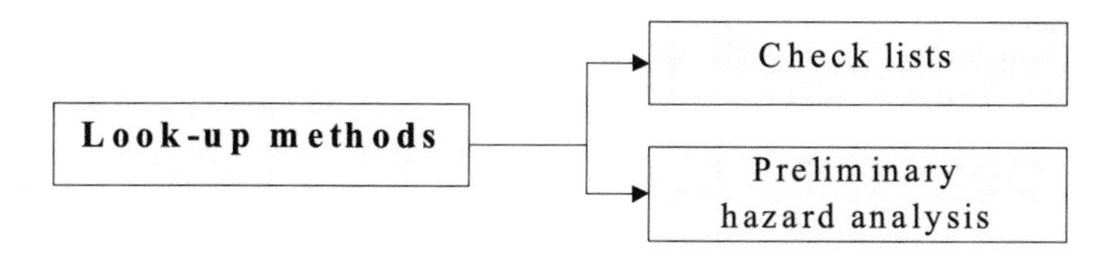

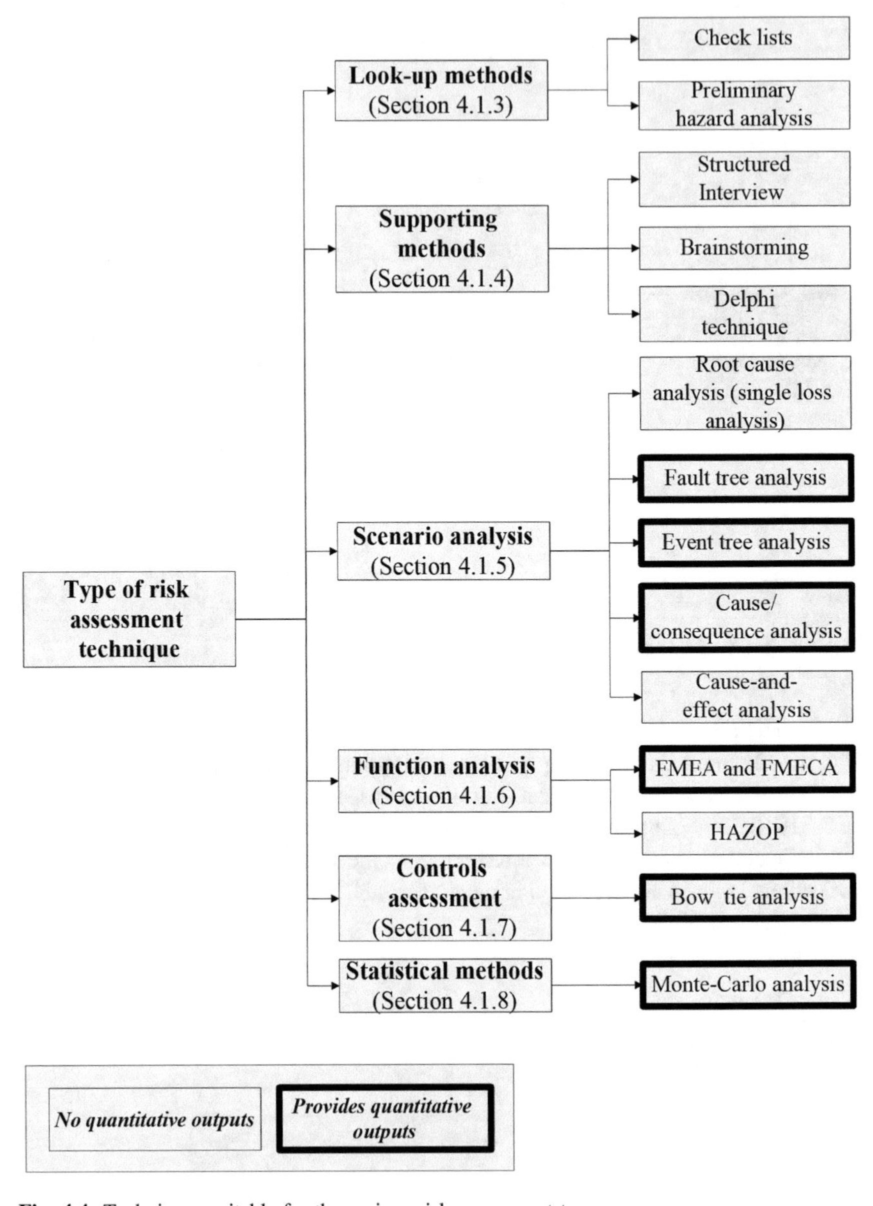

Fig. 4.4 Techniques suitable for the various risk assessment types

Table 4.1 Applicability of tools used for risk assessment (based on ISO/IEC 31010:2009, Annex A, Table A)

Tools and techniques	Sections	Risk assessment process					See Annex A of ISO
		Risk identification	Risk analysis			Risk evaluation	
			Consequence	Probability	Level of risk		
1A	1B	2	3	4	5	6	7
Brainstorming	4.1.4.2	++	o	o	o	o	B 01
Structured or semi-structured interviews	4.1.4.1	++	o	o	o	o	B 02
Delphi simulation technique	4.1.4.3	++	o	o	o	o	B 03
Checklist analysis	4.1.3.1	++	o	o	o	o	B 04
Preliminary hazard analysis	4.1.3.2	++	o	o	o	o	B 05
Hazard and operability studies (HAZOP)	4.1.6.3	++	++	+	+	+	B 06
Hazard Analysis and Critical Control Points (HACCP)		++	++	o	o	++	B 07
Environmental risk assessment		++	++	++	++	++	B 08
Structure « What if? » (SWIFT)		++	++	++	++	++	B 09
Scenario analysis		++	++	+	+	+	B 10
Business impact analysis		+	++	+	+	+	B 11
Root cause analysis	4.1.5.1	o	++	++	++	++	B 12
Failure mode effect analysis (FMEA and FMECA)	4.1.6.1 and 4.1.6.2	++	++	++	++	++	B 13
Fault tree analysis (FTA)	4.1.5.2	+	o	++	+	+	B 14
Event tree analysis	4.1.5.3	+	++	+	+	o	B 15
Cause and consequence analysis		+	++	++	+	+	B 16
Cause-and-effect analysis		++	++	o	o	o	B 17

(continued)

Table 4.1 (continued)

Tools and techniques	Sections	Risk assessment process					See Annex A of ISO
		Risk identification	Risk analysis			Risk evaluation	
			Consequence	Probability	Level of risk		
1A	1B	2	3	4	5	6	7
Layer protection analysis (LOPA)		+	++	+	+	o	B 18
Decision tree	4.1.9.1	o	++	++	+	+	B 19
Human reliability analysis		++	++	++	++	+	B 20
Bow tie analysis	4.1.7.1	o	+	++	++	+	B 21
Reliability centred maintenance		++	++	++	++	++	B 22
Sneak circuit analysis		+	o	o	o	o	B 23
Markov analysis		+	++	o	o	o	B 24
Monte Carlo simulation	4.1.8.1	o	o	o	o	++	B 25
Bayesian statistics and Bayes Nets		o	++	o	o	++	B 26
FN curves		+	++	++	+	++	B 27
Risk indices		+	++	++	+	++	B 28
Consequence/probability matrix		++	++	++	++	+	B 29
Cost/benefit analysis	4.1.9.2	+	++	+	+	+	B 30
Multi-criteria decision analysis (MCDA)		+	++	+	++	+	B 31

++: Strongly applicable
+: Applicable
◦: Not applicable

4.1.3.1 Checklist Analysis

Risk identification checklists are lists of hazards, risks or control failures developed based on historical information and knowledge that has been accumulated from previous similar cases and other sources of information, either as a result of a previous risk assessment or as a result of past failures.

4.1.3.2 Preliminary Hazard Analysis

The Preliminary Hazard Analysis identifies safety-critical areas, provides an initial assessment of hazards and identifies requisite hazard controls and follow-on actions. The Preliminary Hazard Analysis is used to obtain an initial risk assessment of the system hazards.

The preliminary hazard analysis (PHA) technique is a broad, initial study used in the early stages of system design. It focuses on

1. identifying apparent hazards,
2. assessing the severity of potential accidents that could occur involving the hazards, and
3. identifying safeguards for reducing the risks associated with the hazards. This technique focuses on identifying weaknesses early in the life of a system, thus saving time and money that might be required for major redesign if the hazards were discovered at a later date.

PHA relies on brainstorming and expert judgment to assess the significance of hazards and assign a ranking to each situation. This helps in prioritizing recommendations for reducing risks. It is applicable to any activity or system and can be used as a high-level analysis early in the life of a process. It generates qualitative descriptions of the hazards related to a process. Provides a qualitative ranking of the hazardous situations; this ranking can be used to prioritize recommendations for reducing or eliminating hazards in subsequent phases of the life cycle.

Quality of the evaluation depends on the quality and availability of documentation, the training of the review team leader with respect to the various analysis techniques employed, and the experience of the review teams.

PHA focuses predominantly on identifying and classifying hazards rather than evaluating them in detail. It is most often conducted early in the development of an activity or system, when there is little detailed information, or there are few operating procedures. Often a precursor to further risk assessment.

Next, an example is provided of a completed PHA table (Fig. 4.5) documenting the findings of an analysis team.

System	Hazard Description	Risk before Mitigation Measures			Risk Elimination or Mitigation Measures	Risk after Mitigation Measures		
		Severity	Likeli-hood	Risk Hazard Index		Severity	Likeli-hood	Risk Hazard Index
Signalling	Data interface between sub-systmems	Cata-strophic	Probable	20	....	Cata-strophic	Rare	5

Fig. 4.5 Example of a completed PHA table

4.1.4 Supporting Methods

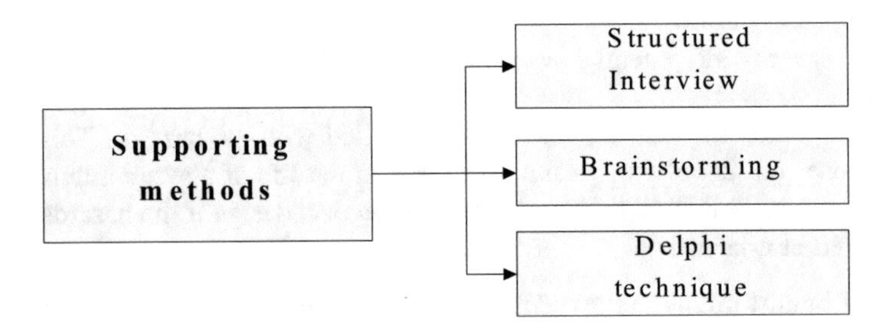

4.1.4.1 Interviewing (Structured or Semi-structured)

In this technique, people with previous experience in similar cases to yours or those with specialized knowledge or industry expertise are interviewed. For a structured interview, a set of questions is prepared in advance and individual interviewees are asked to tell you about any risks that they've experienced or that they think might happen on your case. It encourages the interviewee to view a situation from a different perspective and thus identify risks from that perspective. A semi-structured interview is more a conversation, allows more freedom and aims to explore issues which could arise.

4.1.4.2 Brainstorming

Brainstorming is a way to expand your thinking on a topic. Brainstorming is a technique for gathering ideas, typically, from a group, to identify potential failure modes and associated hazards, risks, criteria for decisions and/or options for treatment. The group could be subject matter experts, team members, risk management team members, and anyone else who might benefit the process of the risk identification process. The technique is to ask them to start identifying possible risk events. The idea behind brainstorming is that one person's idea might spawn another idea, and so on so that by the end of the session you've identified all the possible risks.

Brainstorming can be used as a standalone technique or in combination with other risk assessment methods described briefly next or in detail in the ISO 31010. It can be used at any stage of the risk management process and any stage of the life cycle of a system.

4.1.4.3 The Delphi Method

The name "Delphi Method" refers to the Oracle of Delphi, a priestess at a temple of Apollo in ancient Greece known for her prophecies.

The Delphi method allows experts to work towards a convergent solution, a mutual agreement to a specific problem, by conducting a circulating series of questionnaires and releasing related feedback to further the discussion with each subsequent round. The experts' responses shift as rounds are completed based on the information brought forth by other experts participating in the analysis. The first and very important step is to select a panel of individuals who have experience in the area at issue, they may be from both inside and outside the Organization. It is recommended, the panel members should not know each other, and the process should be conducted with members being at separate locations.

Delphi members are given a questionnaire to identify potential risks. They, in turn, send their responses back to the facilitator of this process. All the responses are organized by content and sent back to the experts for further input, additions, or comments, who then send their comments back one more time, and the facilitator elaborates a final list of risks. The process is continued until group responses converge to a specific solution.

The Delphi technique is a lot like brainstorming, only the people participating in the meeting do not necessarily know each other, and as mentioned, the people participating in this technique can be located at different places and usually participate anonymously. Emails can be used to facilitate the Delphi technique. The Delphi technique is a great tool that allows consensus to be reached quickly. One advantage is that it prevents persons to be unduly influenced by others in the group and in this way it is preventing bias in the outcome because the Delphi members usually do not know each other and also they usually do not know how others in the group responded.

4.1.5 Scenario Analysis

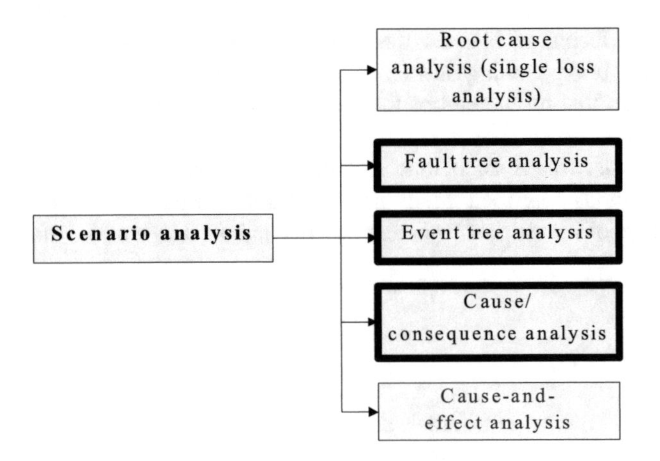

Techniques providing quantitative outputs, are marked with a bold frame (*ISO/IEC 31010:2009).*

4.1.5.1 Root Cause Analysis

Root cause analysis (RCA) (Fig. 4.6) is a method of problem-solving used for identifying the root causes of faults or problems. A factor is considered a root cause if removal thereof from the problem-fault-sequence prevents the final undesirable outcome from recurring; whereas a causal factor is one that affects an event's outcome but is not a root cause. Though removing a causal factor an outcome can benefit, but it does not prevent its recurrence with certainty.

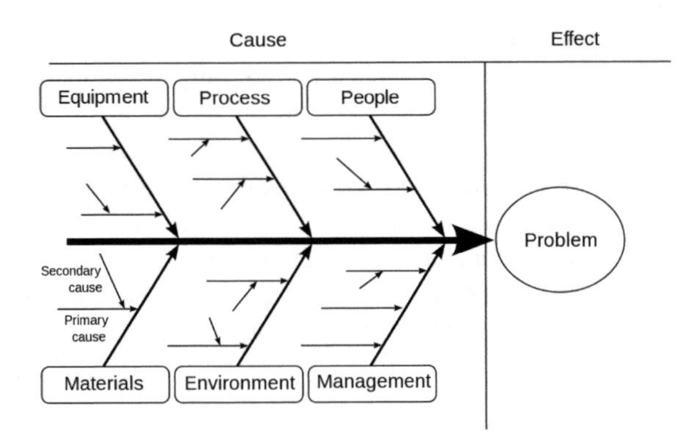

Fig. 4.6 Root cause analysis (RCA)

Essentially it is based on four general principles:

- Define and describe properly the event or problem ('five whys' technique).
- Establish a timeline from a normal situation until the final crisis or failure.
- Distinguish between root causes and causal factor.
- Once implemented (and with constant execution), RCA is transformed into a method of problem prediction.

The diagrams used with this type of analysis are sometimes known as fishbone diagrams because they look like the skeleton of a fish. The technique was developed by Professor Ishikawa in the 1960s.

4.1.5.2 Fault Tree Analysis (FTA)

Fault Tree Analysis (FTA) is a technique for identifying and analysing factors that can contribute to a specified undesired event (called the "top event"). It can be used to predict the most likely failure in a system breakdown (Fig. 4.7).

Fault Tree Analysis may be used qualitatively to identify potential causes and pathways to a failure (the top event) or quantitatively to calculate the probability of the top event, given knowledge of the probabilities of causal events.

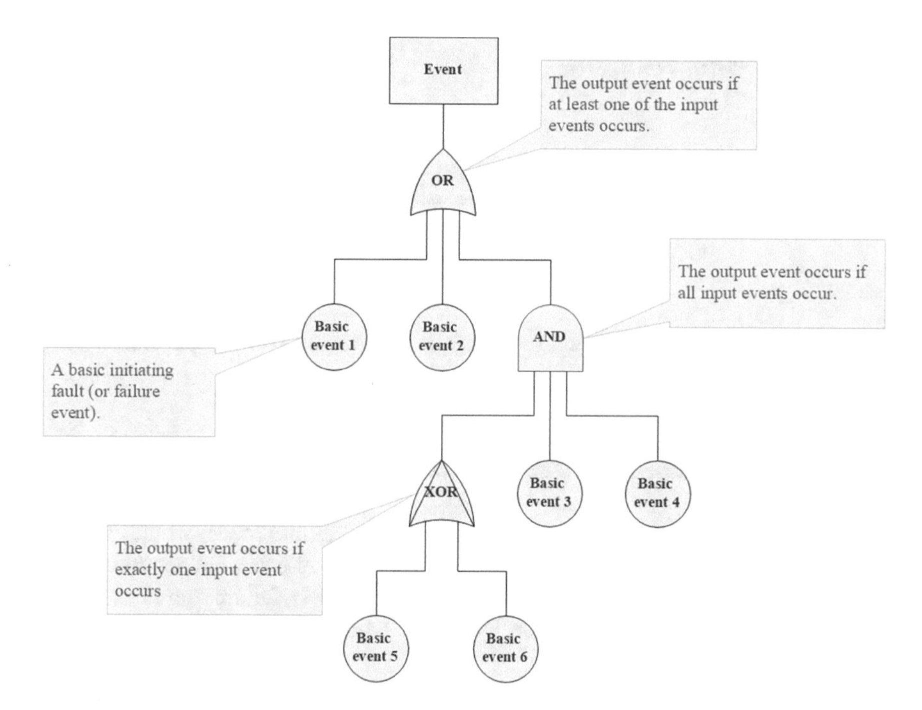

Fig. 4.7 Fault Tree Analysis (FTA)

This analysis method is mainly used in the field of safety engineering and reliability engineering to determine the probability of a safety accident or a particular system level (functional) failure.

This technique is presented in EN 61025:2007 ("Fault tree analysis [FTA]").

4.1.5.3 Even Tree Analysis (ETA)

This analysis technique is used to analyse the effects of functioning or failed systems given that an event has occurred. ETA is a powerful tool that will identify all consequences of a system that have a probability of occurring after an initiating event that can be applied to a wide range of systems (Fig. 4.8).

This technique may be applied to a system early in the design process to identify potential issues that may arise, rather than correcting the issues after they occur. With this forward logic process, use of ETA as a tool in risk assessment can help to prevent negative outcomes from occurring, by providing a risk assessor with the probability of occurrence. ETA uses a type of modelling technique called event tree, which branches events from one single event using Boolean logic.

This technique is presented in EN 62502:2011 ("*Analysis techniques for dependability. Event tree analysis (ETA)*").

4.1.6 Function Analysis

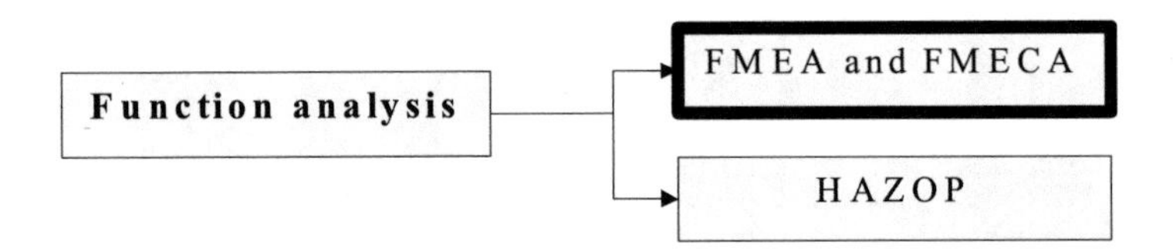

Techniques providing quantitative outputs are marked with a bold frame (*ISO/IEC 31010:2009*).

4.1.6.1 Failure Mode and Effects Analysis (FMEA)

FMEA is a structured process to identify the potential failure modes of the elements of a system, the causes of these failures, and their effects. Failure modes are identified for each component, and the effects of each failure mode on larger assemblies and the whole system are identified. Some of these effects may include hazards.

Potential failure modes can be identified based on past experience with similar products or processes, enabling the team to design those failures out of the system

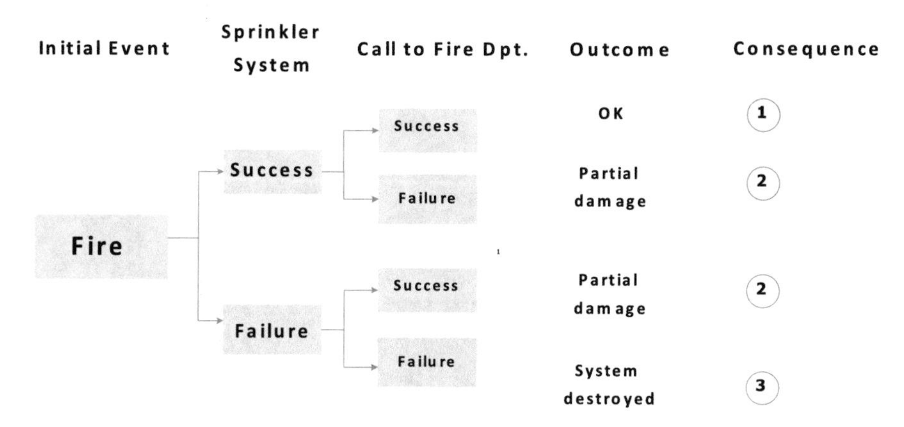

Fig. 4.8 Event Tree analysis of a fire—a simple example

with the minimum of effort and resource expenditure, thereby reducing development time and costs.

Every product or process is subject to different types or modes of failure and the potential failures all have consequences or effects.

- Identify the potential failures and the associated relative risks designed into a product or process
- Prioritize action plans to reduce those potential failures with the highest relative risk
- Track and evaluate the results of the action plans

An example is provided in the next Fig. 4.9.

Column 1:	What is the component or process?
Column 2:	What is the intended function(s)?
Column 3:	A potential failure mode represents any manner in which the component or process step could fail to perform its intended function or functions (i.e. rail may crack/break)
Column 4:	What are the potential causes of failure?
Column 5:	What is the effect(s) if the component or the process fails?
Column 6:	on a scale 1–10 rate the likelihood of each failure (10 = max)
Column 7:	on a scale 1–10 rate the severity of each failure (10 = max)
Column 8:	on a scale 1–10 rate the detectability of each failure (10 = least detectable (Very rare likelihood of detecting failure mode)/ 1 = Almost certain detection of failure mode)
Column 9:	RiskRisk Priority Number (RPN) is the combined weight of Likelihood, Severity and Detectability
Column 10:	Recommended Corrective Action to mitigate the risks
Columns 11–14:	new likelihood, severity and detectability, after recommended actions

Component / Process	Component Function	Failure Mode(s)	Cause(s) Of Failure	Effect(s) Of Failure	Likelihood (L)	Severity (S)	Detection Index (D)	RPN (Risk Priority Number) (L)*(S)*(D)
1	2	3	4	5	6	7	8	9

Recommended Action	After Actions Taken			
	new Likelihood (L)	new Severity (S)	new Detection Index (D)	new RPN (L)*(S)*(D)
10	11	12	13	14

Fig. 4.9 A template for the "Failure Mode and Effects Analysis (FMEA)"

For further guidance see "*EN 60812:2006—Analysis techniques for system reliability. Procedure for failure mode and effects analysis (FMEA)*".

4.1.6.2 Failure Mode, Effects and Criticality Analysis (FMECA)

Failure Mode, Effects and Criticality Analysis (FMECA) is an extension of FMEA by including a *criticality analysis*, which is used to chart the probability of failure modes against the severity of their consequences. The result highlights failure modes with relatively high probability and severity of consequences, allowing remedial effort to be directed where it will produce the greatest value.

4.1.6.3 A Hazard and Operability Study (HAZOP)

A structured and systematic examination of a planned or existing process or operation in order to identify and evaluate problems that may represent risks to personnel or equipment or prevent efficient operation. A HAZOP is a qualitative technique based on guide-words and is carried out by a multi-disciplinary team (HAZOP team) during a set of meetings, to assess the hazard potential that arises from deviation in design specifications and the consequential effects on the facilities as a whole.

This technique is usually performed using a set of guide words: NO/NOT, MORE/LESS OF, AS WELL AS, PART OF REVERSE, AND OTHER THAN. From these guidewords, scenarios that may result in a hazard or an operational problem are identified. Consider the possible flow problems in a process line, the guide word MORE OF will correspond to high flow rate, while that for LESS THAN, low flow rate. The consequences of the hazard and measures to reduce the frequency with which the hazard will occur are then discussed.

This technique had gained wide acceptance in process industries as an effective tool for plant safety and operability improvements.

For further guidance, see EN 61882:2016—Hazard and operability studies (HAZOP studies). Application guide

4.1.7 Controls Assessment

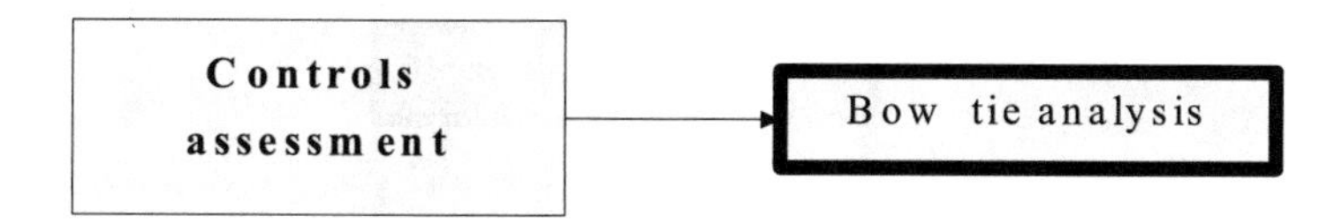

Techniques providing quantitative outputs are marked with a bold frame (*ISO/IEC 31010:2009).*

4.1.7.1 Bow Tie Analysis

The method takes its name from the shape of the diagram that you create, which looks like a men's bowtie (Fig. 4.10).

Bow tie is using a graphical representation and is describing and analysing the pathways of a risk from causes to consequences. It combines the fault tree analysing (cause of an event represented by the knot of a bow tie) and the event tree analysis (analysing the consequences).

A Bowtie diagram is also identifying control measures an Organization has to take to treat the risks.

Once the control measures are identified, the Bowtie method takes it one step further and identifies the ways in which control measures fail.

Besides the basic Bowtie diagram, management systems should also be considered and integrated with the Bowtie. Integrating the management system in a Bowtie demonstrates how An Organization manages hazards. The Bowtie can also be used effectively to assure that Hazards are managed to an acceptable level (ALARP).

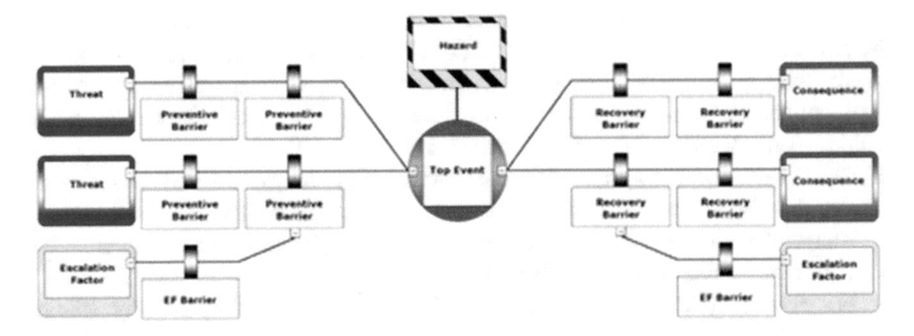

Fig. 4.10 Bowtie diagram

4.1.8 Statistical Methods

Techniques providing quantitative outputs are marked with a bold frame (*ISO/IEC 31010:2009*).

4.1.8.1 Monte Carlo Simulation

Scientists working on the atom bomb first used the technique; it was named for Monte Carlo, the Monaco resort town made famous by its casinos. With games of chance, all the possible outcomes and probabilities are known, but the set of future outcomes is unknown. It is up to the analyst to determine the set of outcomes and the probability that they will occur. In Monte Carlo simulation, the analyst runs multiple trials (often thousands) to determine all the possible outcomes and the probability that they will take place. It lets you see all the possible outcomes of your decisions and assess the impact of risk, allowing for better decision making under uncertainty. The essential idea is using randomness to solve problems that might be deterministic in principle. In principle, Monte Carlo methods can be used to solve any problem having a probabilistic interpretation.

Monte Carlo model approximates solutions to quantitative problems through statistical sampling.

It is a decision-making tool that integrates the concept that every decision will have some impact on overall risk. The probability distributions produced by a Monte Carlo model create a picture of risk. Because of advances in software, very complex Monte Carlo models can be designed and executed by anyone with access to a personal computer.

Since its introduction in World War II, Monte Carlo simulation has been used to model a variety of physical and conceptual systems.

4.1.9 Other Techniques

4.1.9.1 Decision Tree Analysis

A Decision Tree Analysis is a graphic representation of various alternative solutions that are available to solve a problem. It is a diagram that shows the implications of choosing one or other alternatives. The manner of illustrating often proves to be decisive when making a choice. Because each decision or event node has at least two alternatives, the structure of the decision looks like a tree, typically placed on its side with the root on the left and the branches on the right, with potentially many branches.

4.1.9.2 Cost-Benefit Analysis

Introduction

Cost-benefit analysis (CBA) is a useful tool for organizing, assessing and finally presenting the cost and benefits, and pros and cons of interventions [1]. A CBA allows comparisons between all the possible alternatives to assist the decision-makers in examining the most profitable safety measure to invest.

Risk is commonly defined as the probability of potential impacts affecting people, assets or the environment. Natural disasters may cause a variety of effects which are usually classified into social, economic, and environmental impacts as well as according to whether they are triggered directly by the event or occur over time as indirect or macroeconomic effects (Fig. 4.11).

Two important issues deserve special attention when conducting a CBA [1].

1. Assessment of risk: The analysis should be done by analyses that should take account of the probability of future disaster events occurring (stochastic manner), in order to account for the specific nature of natural hazards and associated disaster impacts.

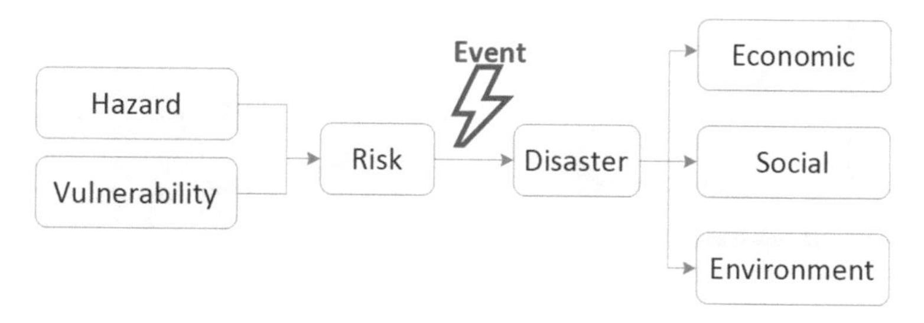

Fig. 4.11 Disaster, risk and categories of potential disaster impacts (as per [1])

2. Assessment of avoided risks: As disaster risk is a downside risk, benefits are the risks avoided. The core benefit generated by investments in disaster risk management is the reduction in future impacts and losses.

Estimating the economic efficiency of an intervention, benefits and costs need to be compared. Costs and benefits arising over time need to be discounted to render current and future effects comparable. From an economic point of view, $1 today has more value than $1 in 10 years; thus future values need to be discounted by a discount rate representing the preference for the present over the future. Furthermore, costs and benefits are compared under a common economic efficiency decision criterion to assess whether benefits exceed costs.

Cost-benefit analysis (CBA) provides an objective means of comparing the costs and benefits of the risk without treatment and the comparable costs and benefits of the treated risk (Fig. 4.12).

There should be a consistent approach to comparing the costs and benefits of different options. All the benefits should be considered: both direct benefits and indirect benefits as also both direct and indirect costs. Costs and benefits may be quantitative or qualitative.

Benefits can arise:

a. Directly from the reduction in risk.
b. Increased opportunities.
c. Indirectly such as from greater management confidence, savings such as insurance premium reductions, or improvements in intangibles like reputation or credit rating.

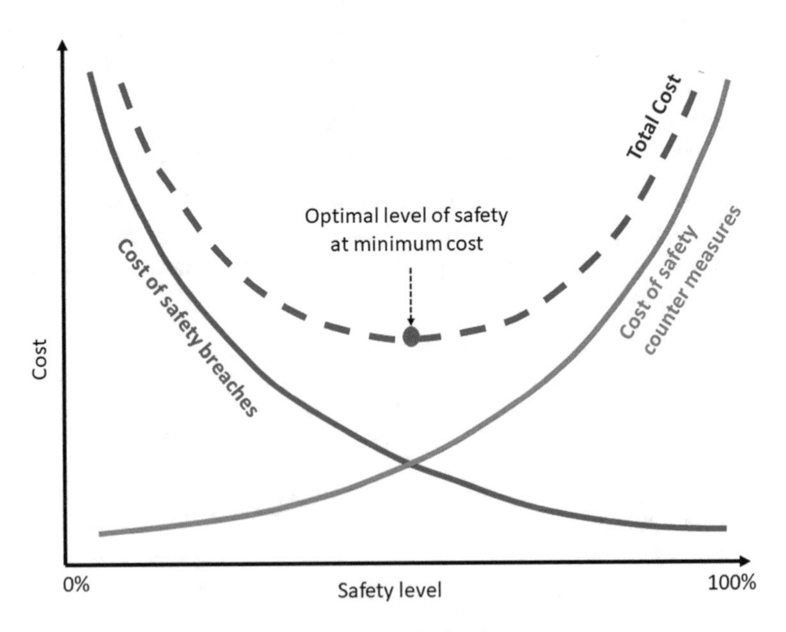

Fig. 4.12 Relation between the level of safety and related cost

Costs can be:

a. Direct costs related with treatment options and their implementation.
b. Increased risk of negative outcomes or reduced opportunities.
c. Indirect costs such as loss of productivity, disruption from core business activities, management time, etc.

Three decision criteria are of significant importance in CBA [2]:

- See section **Net Present Value (NPV)**.
- See section **Cost Benefit Ratio**.
- **Internal Rate of Return (IRR)**: Whereas the above two criteria use a fixed discount rate, this criterion calculates the interest rate internally, which represents the return on investments in the given project. A project is rated desirable if this IRR surpasses the average return of public capital determined beforehand (i.e. 12%).

The Net Present Value (NPV)

The NPV is the most useful and one of the most commonly used criteria for determining whether an intervention should be accepted. The net present value formula is:

$$\text{NPV} = \sum_{t=0}^{n} \frac{(B_t - C_t)}{(1 + r)^t}$$

where,

B_t are benefits in period t
C_t are costs in period t
r is the appropriate financial or economic discount rate
n is the number of years for which the project will operate

For input to our calculation, future costs are converted to their current equivalent by using a suitable discount rate: in the analysis of the interventions, any costs and benefits of an intervention that are received in future periods are discounted, or deflated by some factor, r. The factor used to discount future costs and benefits is called the discount rate and is usually expressed as a percentage.

Example:

100 USD receivable today is more than 100 USD receivable a year later, as 1000 USD received today will earn interest or profits and shall accumulate to more than 100 USD in a year. Alternatively, 100 USD received today can be used to reduce borrowing thereby avoiding interest payments as well as reducing debts by 100 USD. Assuming that the Railways' cost of finance is its current dividend rate (say 6% per year), USD 106 received a year hence should be worth 100 USD today and 100 0USD which may be received in a year is worth about 94 USD today (actually it is worth 94.34 USD). Likewise, the present value of 100 USD receivable 2 years hence is about 89 USD, and so on. In this way, the cash flow for the intervention in any future year can be discounted to obtain the present value.

For example, suppose an intervention is expected to yield a stream of benefits equal to $B_0, B_1, B_2, \ldots, B_n$ and to incur a stream of costs equal to $C_0, C_1, C_2, \ldots, C_n$ in years 0, 1, 2, …, n. Then in each period, the net benefits (benefits minus costs) of the project will be:

$(B_0–C_0), (B_1–C_1), (B_2–C_2), \ldots, (B_n–C_n)$

This is simply the intervention's net benefit flow.

If the discount rate, r, is constant, then the discounted cash flow of the project can be represented as:

Year	Benefit flow
Year 0	$B_0–C_0$
Year 1	$\frac{(B_1 - C_1)}{(1+r)}$
Year 2	$\frac{(B_1 - C_1)}{(1+r)^2}$
Year 3	$\frac{(B_1 - C_1)}{(1+r)^3}$
Year n	$\frac{(B_1 - C_1)}{(1+r)^n}$

Once future net income streams have been discounted in this way, expenditures and revenues from all the different time periods will be valued in units of similar value—present day units of currency. They will then be directly comparable with each other and can be added together. Adding the discounted net benefits from each year of the intervention, life, its discounted net benefit flow, gives a single monetary value called the intervention's net present value, NPV. For, the previous example, the intervention's NPV is:

The net present value criterion of an intervention is the single most important measure of the intervention's worth.

In the above Table 4.2, an r = 6% discount rate is used to discount the net benefits of a railway intervention. The intervention's NPV can then be estimated by just adding up these discounted net benefits. Columns (1), (2) and (3) show the non-discounted costs, benefits and net benefits (benefits-costs) of the railway intervention.

Column (4) gives the discount factor, $1/(1+r)^t$, by which the non-discounted net benefits in column (3) are multiplied, to obtain the discounted value of these net benefits in each year, t, shown in column (5). These discounted net benefits can then be added together to obtain the total discounted net benefits, or net present value, of the intervention.

The bottom line of the table shows that the NPV comes to 57.06 million USD if a 6% discount rate is used. An NPV higher than zero indicates that the discounted benefits of the intervention are expected to be higher than its discounted costs and the intervention will, therefore, be worth undertaking.

Table 4.2 Example: Cash flow of an intervention—discounted at 6% discount rate (million USD)

Year (t)	Costs (1)	Benefits (B) (2)	Net Benefits (3)=(2)-(1)	Discount Factor $1/(1+r)^t$ (4)	Net Benefits (5)=(3)*(4)
0	100	0	-100	1	-100,00
1	400	50	-350	0,943	-330,19
2	200	150	-50	0,890	-44,50
3	100	200	100	0,840	83,96
4	100	200	100	0,792	79,21
5	100	200	100	0,747	74,73
6	100	200	100	0,705	70,50
7	100	200	100	0,665	66,51
8	100	350	250	0,627	156,85
Total	**1.300**	**1.550**	**250**	**NPV =**	**57,06**

This example illustrates how crucially the estimation of an intervention's NPV depends on the discount rate employed.[1] A lower discount rate would have deflated future income by less and increased NPV of the intervention. A higher discount rate would have deflated future income more heavily and decreased the NPV of the intervention, possibly changing it from positive to negative. The selection of the appropriate discount rate is, therefore, a critical issue in intervention appraisal.

In the above example, a discount rate of $r = 7\%$ gives an NPV of 32.85 Million USD, and a discount rate of $r = 8\%$ gives an NPV of 10.44 Million USD. A discount rate of $r = 8.5\%$ gives a negative NPV (−0.15 Million USD).

The NPV of an intervention is –as presented- the sum of the present values of the net cash flows for all the years of the intervention's economic life (present value of incomes minus present value of expenses). Interventions and processes with the

[1]The discount rate is roughly the opportunity cost of capital: it is the cost of using the capital in one project renouncing to earn a return in another project. Its value is defined mostly empirically for a given project, in a given country or region, for a given firm and at a given time. The value of the discount rate can have a very serious impact on the decision making process of a cost benefit or life cycle cost analysis.

highest NPV are usually the winners. Often incremental changes on an intervention can lead to a positive NPV. Thus many improvement interventions must be selected on the least negative NPV values from many alternatives.

NPV in decision making:

If ...	**It means ...**	**Then ...**
NPV > 0	the investment on the planned intervention would add value to the Infrastructure Manager	the intervention may be accepted
NPV < 0	the investment on the planned intervention would subtract value from the Infrastructure Manager's or government's available budget	the intervention should be rejected
NPV = 0	the investment on the planned intervention would neither gain nor lose value for the Infrastructure Manager's or government's available budget	We should be indifferent in the decision whether to accept or reject the intervention. This intervention adds no monetary value A decision should be based on other criteria, i.e. strategic positioning or other factors not explicitly included in the calculation

The Cost-Benefit Ratio

The Cost-Benefit Ratio is a variant of the NPV. The benefits are divided by the costs. If the ratio is higher than 1, i.e. benefits exceed costs, a project is considered to add value to society.

Costs and benefits should be calculated over an appropriate time span, on the basis of discounted cash flow.

$$Cost\,Benefit\,Ratio = \frac{Net\,Present\,Value\,of\,Benefits}{Net\,Present\,Value\,of\,Costs} > 1$$

Benefits =

- value of avoided injuries+
- damage avoided+
- other benefits.

Costs should be shared by those who benefit from the reduction of the risk.

Qualitative Analysis of Costs and Benefits

Cost-benefit analysis (CBA) presented in Sects. 4.1.9, and 4.1.9.2, is comparing estimated costs and benefits. In many cases, it will not be possible to quantify all costs and all benefits and sometimes benefits cannot be quantified at all. For example, preventing the damage to reputation caused by a major incident cannot be easily quantified.

Cost-Benefit of Risk Reduction

Costs and benefits should be calculated over an appropriate time span, on the basis of discounted cash flow.

$$\textit{Cost Benefit Ratio} = \frac{\textit{Net Present Value of Benefits}}{\textit{Net Present Value of Costs}}$$

Benefits =

- value of avoided injuries+
- damage avoided+
- other benefits.

Costs should be shared by those who benefit from the reduction of the risk.

Value of Avoided Deaths and Injuries

The cash valuations of preventing health and safety effects on people are presented for UK (2003) and New Zealand (2017)

United Kingdom (http://www.hse.gov.uk/risk/theory/alarpcheck.htm)

		Values in £	Values in USD[2]
Fatality		£1,336,800 (times 2 for cancer)	$ 1,690,000
Injury			
Permanently incapacitating injury	Moderate to severe pain for 1–4 weeks. Thereafter some pain gradually reducing but may recur when taking part in some activities. Some permanent restrictions to leisure and possibly some work activities	£207,200	$ 262,000
Serious	Slight to moderate pain for 2–7 days. Thereafter some pain/discomfort for	£20,500	$ 25,900

(continued)

[2]Approximately, as per June 2019.

(continued)

	several weeks. Some restrictions to work and/or leisure activities for several weeks/months. After 3–4 months, return to normal health with no permanent disability		
Slight	Injury involving minor cuts and bruises with a quick and complete recovery	£300	$ 380
Illness			
Permanently incapacitating illness	Same as for injury	£193,100	$ 244,000
Other cases of ill health	Over one-week absence. No permanent health consequences	£2300 + £180 per day of absence	$2900 + $230 per day of absence
Minor	Up to one-week absence. No permanent health consequences	£53	$ 67

New Zealand

As a guide the value of avoided deaths and injuries can be taken as [3]:

Injury (2017)	Value (2017) (New Zealand Dollars)	Value (USD) (Approximately, as per June 2019)
Fatality	$ 4,915,000	$ 3,212,000
Serious injury	$ 513,000	$ 335,000
Minor injury	$ 29,000	$ 19,000

Example of CBA Calculation

Case of serious train/car accident because passing level crossing barriers Example of the case examined "Serious train/car accident because of passing level crossing barriers." (Figure 4.13)

1. We estimate the probability of the accident to happen 40.2%
2. Let us for this example assume that if an accident happens,
   - 1 person will die,
   - persons will be seriously injured and
   - persons minor injured
   - 3 cars will be damaged
3. The value will be estimated:
   - 1 person died X 1,000,000 USD = 1,000,000 USD
   - 2 persons will be seriously injured = 2 × 100,000 = 200,000 USD

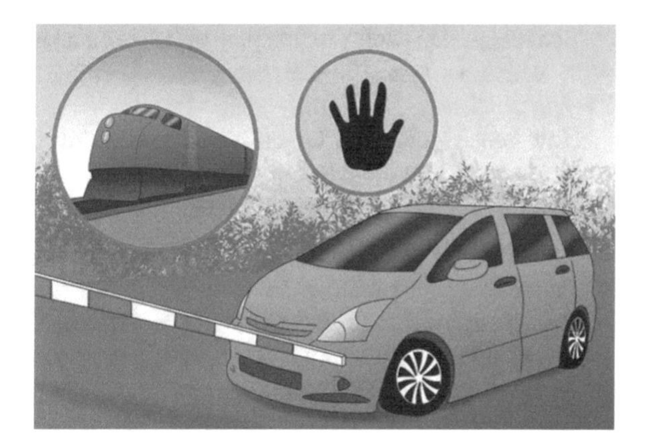

Fig. 4.13 Level crossing

- 3 persons minor injured = 3 X 5,000 = 15,000 USD
- Car damage = 25,000 USD

TOTAL = 1,240,000 USD

4. We plan to improve the safety of the road crossing, with a cost estimated on 80,000 USD.
5. We calculate that the probability of an accident will decrease from 40.2 to 25%.
6. What will be the Cost-Benefit of Risk Reduction?
7. As we discussed: $Cost\,Benefit\,Ratio = \frac{Net\,Present\,Value\,of\,Benefits}{Net\,Present\,Value\,ofCosts}$
8. We calculate:

- The benefit will be a decrease of the accident probability from 40.2% to 25% = 15.2%
- The *Value of Benefits* will be 15.2% X1, 240,000 USD $\approx$ 190,000 USD
- The *Value of Costs* will be 80,000 USD
- So, the Cos$t\,Benefit\,Ratio = \frac{190{,}000}{80{,}000} \approx 2.4 \gg 1$

So, the investment in improving the safety of the road crossing is extremely beneficial.

Explosion in a rolling stock maintenance depot

A simple method for coarse screening of measures is presented below. This puts the costs and benefits into a common format of 'USDs per year' for the lifetime of a plant.

Consider a rolling stock maintenance depot with a process that if it were to explode could lead to:

- 20 fatalities
- 40 permanently injured
- 100 seriously injured
- 200 slightly injured

The rate of this explosion happening has been analysed to be about 1×10^{-5} per year, which is 1 in 100,000 per year. The plant has an estimated lifetime of 25 years.

How much could the Organization reasonably spend to eliminate (reduce to zero) the risk from the explosion?

If the risk of explosion were to be eliminated the benefits can be assessed to be:

- Fatalities: $20 \times 1{,}336{,}800 \times 1 \times 10^{-5} \times 25$ years = 6684
- Permanent injuries: $40 \times 207{,}200 \times 1 \times 10^{-5} \times 25$ years = 2072
- Serious injuries: $100 \times 20{,}500 \times 1 \times 10^{-5} \times 25$ years = 512
- Slight Injuries: $200 \times 300 \times 1 \times 10^{-5} \times 25$ years = 15
- Total benefits: USD 9.283

The sum of USD 9.283 is the estimated benefit of eliminating the major accident explosion at the plant on the basis of avoidance of casualties. (This method does not include discounting or take account of inflation.)

For a measure to be deemed not reasonably practicable, the cost has to be grossly disproportionate to the benefits. This is taken into account by the disproportion factor (DF). In this case, the DF will reflect that the consequences of such explosions are high. A DF of more than 10 is unlikely.

Therefore it might be reasonably practicable to spend up to somewhere in the region of USD 93,000 (USD 9300 $\times$ 10) to eliminate the risk of an explosion. The duty holder would have to justify the use of a smaller DF.

This type of simple analysis can be used to eliminate or include some measures by costing various alternative methods of eliminating or reducing risks.

4.1.9.3 Other Techniques not Mentioned in the ISO/IEC 31010:2009

Next, other techniques not mentioned in the ISO/IEC 31010:2009 are briefly presented.

Three-Point Estimate

Three-Points estimation is a technique that involves people that are professional in the task we are estimating by this technique.

It is called three-point estimation because the team members provide their pessimistic, optimistic and best guess estimates for their risk estimation, based on prior experience or best-guesses.

Three-point estimation is a:

- Triangular distribution (Simple Average)
- Beta distribution (Weighted Average).

The process for the Tree Point Estimation technique

Team members involved in the process are requested to make three estimates: the pessimistic (P), the most likely (M) and the optimistic (O) estimation. Then you do some simple mathematics with the three estimates:

Three-point Estimation[3]: $\frac{P + 4M + 0}{6}$

Standard deviation: $\frac{P - 0}{6}$

The calculation reflects the amount of risk in the task and the severity of the impact of optimistic and pessimistic risks.

Standard deviation is the possible range for the estimate. **You can assess and compare the risk of various cases by looking at the ranges of the cases and the standard deviations.**

Expected Monetary Value (EMV)

Expected Monetary Value (EMV) is a method used to establish the contingency reserves for a project budget and schedule.

As we discussed, once you have identified your risks, you need to calculate out both the likelihood of the threats being realized, and their possible impact. One way of doing this is to make your best estimate of the probability of the event occurring, and then to multiply this by the amount it will cost you to set things right if it happens. This gives you a value for the risk:

Risk Value = Probability of Event × Cost of Event

Or

If we express the risk value as the Expected Monetary Value (EMV):

EMV = P * I

(P =Probability, I = Impact)

Example

Imagine you have a business and you have identified a risk that your rent might rise substantially.

You think that within the next year there is a 70% chance that this will happen because your landlord has recently increased rents for other businesses.

If this happens, over the next year it will cost your business an additional $350,000.

So the rent increase risk value is: 70% (Probability of Event) × $500,000 (Cost of Event) = $245,000 (Risk Value)

[3]Or "Beta distribution".

Expert Judgment

Risk identification experts can include anyone with experience in working on similar projects, experience working in the business area for which the project was conducted, or industry-specific experience. When using this technique, you should consider any bias that your experts may have with regard to the project or potential risk events.

SWOT Analysis

SWOT Analysis is a useful technique for understanding your Strengths and Weaknesses, and for identifying both the Opportunities open to you and the Threats you face (Fig. 4.14).

In general, strengths and weaknesses are related to issues within the Organization. Strengths examine what's going well with your Organization and what your customers or the marketplace see as your strengths. Weaknesses are areas that may be improved by the Organization. Negative risks are typically associated with the weaknesses of the Organization and its strengths are associated with positive risks. The Organization usually has external opportunities and threats. SWOT analysis is sometimes referred to as internal-external analysis and can be used to help discover and document potential risks in combination with brainstorming techniques.

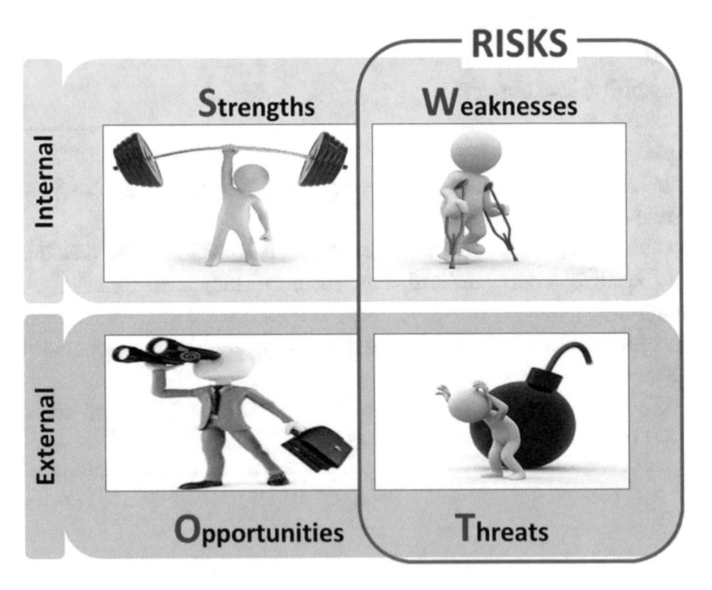

Fig. 4.14 SWOT analysis structure

Sensitivity Analysis

Sensitivity Analysis is a technique used to determine which risks affect a project the most. It is a quantitative method of analysing the potential impact on the project of risk events and determining which risk event (or events) has the highest impact potential by examining all the uncertain elements at their baseline values. A tornado diagram is one way to display sensitivity analysis data.

A Functional Hazard Analysis (FHA)

A systematic, comprehensive examination of functions to identify and classify failure conditions of those functions according to their severity. For the analysis of a change to the railway, it may be appropriate to apply the FHA at the system level. This would involve a high-level, qualitative assessment of the defined functions of the system (as specified in the system definition).

The system-level FHA is undertaken to identify and classify the failure conditions associated with the system-level functions.

FHA involves less work than FMEA/FMECA and can be started earlier, because a specification, and not a design, is all that is required. However, FHA is not good at finding hazards that are not easily characterised as the failure of a function (such as electromagnetic interference or fuel leakage).

Failure Block Diagram Analysis

The analysis of a system based on the component reliabilities.

It is a method of modelling how components and subsystem failures combine to cause system failure. Reliability block diagrams may be analysed to predict the availability of a system and determined the critical components from a reliability viewpoint.

4.1.10 Accident Rates

Fatality rates can be expressed in the following terms [4], see also Sect. 3.5.4.3:

Fatality Accident Rate (FAR)

The FAR or Fatal Accident Rate is a measure of how many people would die per 100 million exposure hours. This is approximately the same as saying how many deaths are likely in 1000 people, over their working lives. It assumes an average of working 2000 h a year, and a working life of 50 years (Note, that standard hours worked per annum is 1886 h). By their very nature, FARs vary significantly throughout a passenger trip or working day. An average rate of exposure is therefore used.

$$FAR = \frac{100,000,000 \ * \ DPA\,(Death\,per\,annum)}{(Number\,of\,people\,exposed) \ * \ (House\,exposed\,per\,annum)}$$

Equivalent Fatal Accident Rate (EFAR)

Injuries will be considered as if 10 serious injuries are equivalent to death, and 200 minor injuries are equivalent to death.

Therefore, Equivalent Deaths per annum (EDPA)

$$EDPA = DPA + \left(\frac{serious\,injuries\,p.a.}{10} + \frac{minor\,injuries\,p.a.}{200}\right)$$

where

DPA = deaths per annum

p.a. = per annum

Then Equivalent Fatal Accident Rate (EFAR):

$$EFAR = \frac{100,000,000 \ * \ EDPA}{(Number\,of\,people\,exposed) \ * \ (Hours\,exposed\,p.a.)} \quad (4.5)$$

where

EDPA = Equivalent Death per annum

Often the DPA or EDPA will have to be assessed statistically, i.e. one death may be expected every 20 years giving a likelihood of 0.05 deaths p.a. Where no detailed information such as accident history is available, consideration of any industry-wide information may assist.

Example of Accident Rate calculation

- Say 500 rail personnel in a workgroup, each working 1886 h per annum
- In this workgroup say there have been 3 fatalities in the last 15 years
- From the personal accident database, there have been 6 serious injuries and 113 minor injuries since in the last 12 months

So, DPA = 3/15 = 0.2

which = 0.2 in 500, which = 1 in 2500, which is less than 1 in 1000 so it is below the Upper Bound (see Sects. 3.5.4.2 and 3.5.4.3). $FAR = \frac{100{,}000{,}000 * 0.2}{500 * 1886} =$ 21.2 which is less than 50 (see Sect. 3.5.4.3)

$EDPA = 0.2 + \frac{6}{10} + \frac{113}{200} = 1.365$ in 500

Which is 1 in 366, which is greater than 1 in 400, so it is above the Upper Bound (see Sects. 3.5.4.2 and 3.5.4.3).

$EFAR = \frac{100{,}000{,}000 * 1.365}{500 * 1886} = 145$, which is greater than 125 (see Sect. 3.5.4.3).

Therefore, the FAR is ALARP (or tolerable), using the table of upper and lower bounds in Sect. 3.5.4. The EFAR is intolerable (risk cannot be justified) and therefore, action is required to reduce the risk to ALARP.

4.2 The Risk Management Plan

4.2.1 Introduction

The Risk Management Plan presents at a high level how an Organization manages its risks and how the entire process is integrated into the activities of the Organization.

The Risk Management Plan may contains [2]:

a. A statement of the Organization's risk management policy.
b. A description of the external and internal context, arrangements for corporate governance and supervision, and the environment in which the Organization operates.
c. Details of the scope and objectives of the risk management activities in the Organization, including organizational criteria for assessing whether risks are tolerable.
d. Risk management responsibilities and functions in the Organization.
e. The list of risks identified and an analysis of them, usually in the form of a risk register included as an appendix.
f. Summaries of the risk treatment plans for major risks, incorporated as an Appendix or by reference to a treatment plan.

The risk treatment plan is produced after the Organization has conducted its risk assessment and is a detailed document describing roles and responsibilities for specific actions to bring the identified risks down to an acceptable level. The risk treatment plan needs to provide a summary of (Table 4.3):

Table 4.3 Risk Treatment Plan template

Area/Department		**Risk Register ID**	
Date Treatment Developed		**Risk Category**	
Risk Owner		**Treatment Owner**	

Risk Description	Control Effectiveness	Risk Treatment			Monitor & Review	Implementation Status
		Treatment Action	Responsibility	Implementation Date		
Provide a description of what the risk is.	*Provide the last known control effectiveness rating (i.e. high, medium or low).*	*Selecting the most appropriate treatment option involves balancing the costs and efforts of implementation against the benefits derived. Factors such as legal, regulatory, the natural environment and social corporate responsibility must also be considering when deciding on treatment options.* *Some examples of treatment options are:* *Avoiding the risk by deciding not to start or continue the activity that gives rise to the risk* *Taking or increasing the risk in order to pursue an opportunity* *Removing the risk source* *Changing the likelihood* *Changing the consequence* *Sharing the risk with another party* *Retaining the risk by informed decision*	*Assign a person to authorise the time and resources required for risk treatment.* *Assign person must implement the risk treatment.*	*Agreed timeframes must be establishing to outline when risk treatment will be achieved and completed.*	*Consideration must be given to how risk treatment will be monitored to assess whether the treatment is effective.* *The frequency and method of how progress against treatment plans is reported must be defined.*	*Provide a status with regard to implementation progress (i.e. not started, in progress or completed).*

a. Identified risks
b. Responses that have been designed for each risk
c. Parties responsible for those risks and
d. The date to apply the risk treatment.

4.2.2 The Risk Register

ISO 73:2009 Risk management—Vocabulary [5] defines a risk register to be a "*record of information about identified risks*". A Risk Register (also referred to as a Risk Log), is a master document which is created during the early stages of the process. It is a tool helping you to track issues and address problems as they arise.

The *Risk Register* records details of all the risks identified at the beginning and during the life of projects, their grading in terms of likelihood of occurring and seriousness of impacts, initial plans for mitigating each high-level risk, the costs and responsibilities of the prescribed mitigation strategies and subsequent results.

It usually includes [6]:

- a unique identifier for each risk
- a description of each risk and how it will affect the project
- an assessment of the likelihood it will occur and the possible seriousness/impact if it does occur (low, medium, high)
- a classification of each risk according to a risk assessment table
- who is responsible for managing the risk
- an outline of proposed mitigation actions (preventative and contingency) and
- in larger projects, costings for each mitigation strategy.

This Register should be maintained throughout the project and will change regularly as existing risks are re-graded in the light of the effectiveness of the mitigation strategy, and new risks are identified. In smaller projects, the *Risk Register* is often used as the *Risk Management Plan.*

A Risk Register is developed to:

- provide a useful tool for managing and reducing the risks identified before and during the project
- document risk mitigation strategies being pursued in response to the identified risks and their grading in terms of likelihood and seriousness
- provide the Project Sponsor, Steering Committee/senior management with a documented framework from which risk status can be reported
- ensure the communication of risk management issues to key Stakeholders
- provide a mechanism for seeking and acting on feedback to encourage the involvement of the key Stakeholders and
- identify the mitigation actions required for the implementation of the risk management plan and associated costings.

Initial risks must be identified and classified according to the likelihood and seriousness very early in the Project. This initial risk assessment will form part of the Project Proposal/Brief or Project Business Case for the project. Once the project is approved the Risk Management Plan and Risk Register should be fully developed. In the case of smaller projects, the Risk Register may serve both purposes.

The completed *Risk Register* should be brief and to the point, so it quickly conveys the essential information. It should be updated regularly, at least monthly.

The description of the risk should include the associated consequences or impact where these are not obvious. These consequences can be useful in identifying appropriate mitigation actions. In larger more complex projects, a separate column may be required.

Mitigation actions should include such things as:

- Preventative actions—planned actions to reduce the likelihood a risk will occur and/or reduce the seriousness should it occur. (What should you do now?).
- Contingency actions—planned actions to reduce the immediate seriousness of the risk when it does occur. (What should you do when?)
- Recovery actions—planned actions taken once a risk has occurred to allow you to move on. (What should you do after?).

References

1. Mechler R, The Risk to Resilience Study Team (2008) *The cost-benefit analysis methodology, from risk to resilience working paper No. 1*. In: Moench M, Caspari E, Pokhrel A (eds), ISET, ISET-Nepal and ProVention, Kathmandu, Nepal, 32 pp
2. Standards Australia/Standards New Zealand (2005) Risk Management Guidelines—Companion to AS/NZS 4360:2004
3. New Zealand, Ministry of Transport (2017) *Social Cost of Road Crashes and Injuries June 2017 update*
4. Australian/New Zealand Standard (2004) *AS/NZS 4360 SET Risk Management*
5. ISO/IEC Guide 73, Risk Management—Vocabulary—Guideline for use in standards [3] ISO/ IEC 31010/ 2009, Risk Management—Risk Assessment Techniques
6. Department of Premier and Cabinet (2008) *Project Risk Register Template & Guide*, Version 1.3, April 2008, Tasmania

Chapter 5
Health Risk Management (For Staff with Safety-Critical Positions)

5.1 Introduction

This chapter introduces Health Risk Management to assist rail Organizations in performing health risk assessments for staff with a safety-critical position.

Rail Organizations shall establish systems and procedures to ensure staff with safety-critical positions receive the appropriate level of health assessment as outlined in the national standards and legal frameworks. The systems and procedures should be based on a risk management approach so that the level and frequency of health assessment of staff with a safety-critical position correspond with the risks associated with the tasks they perform. Rail Organizations should perform their risk assessments of work related to safety in their operating environment and apply health assessments accordingly.

Following will be presented:

- an overview of the risk categories and matched health assessment provisions
- a step-by-step guide to the risk assessment of tasks critical to safety and the health attributes needed for the tasks
- guidance for identifying health assessment requirements for specific tasks, such as colour vision.

Work related to safety may also have specific occupational health and safety risks associated with it, for example, noise or manual handling. These are outside the scope of this Chapter and should be managed as required by relevant occupational health and safety regulations.

This chapter is based on the Australian "National Standard for Health Assessment of Staff with a safety-critical position" [1].

K. Tzanakakis, *Managing Risks in the Railway System*, Springer Tracts on Transportation and Traffic 18, https://doi.org/10.1007/978-3-030-66266-0_5

5.2 Definitions Used in This Chapter

Around the Track Personnel	Staff who perform safety non-critical tasks on or near the track
Controlled environment	A rail workplace such as a siding, rail yard or workshop environment where a risk assessment has been performed to identify hazards and where controls are implemented to ensure that any person working in or transiting the area is not placed at risk from moving trains The essential requirement of a Controlled Environment is that it must protect staff from moving rolling stock
Health Professional	Health professional who is appointed and authorised by accredited rail Organizations to conduct health assessments for staff who perform work critical to safety, in line with the related regulations
Rail Organization	Means any of the following: a. a Railway Infrastructure Manager b. a Train Operating Company (Railway Undertaking) c. a Rail Vehicle Owner d. a Maintenance or Inspection Provider e. any other Entity prescribed as a Rail Organization by regulations
Safety-critical Staff	Staff with a safety-critical position
Serious incident	An accident or incident that affects the public or rail network resulting in the death of a person; incapacitating injury to a person; a collision or derailment involving rolling stock that results in significant damage; or any other occurrence that results in significant property damage

5.3 Work Critical to Safety

Work critical to safety is any of the following:

a. Driving a vehicle
b. Dispatching and controlling the movement of vehicles
c. Signalling and signalling operations
d. Operation of level crossing equipment
e. Receiving and relaying communications or other activity capable of controlling the movement of vehicles
f. Coupling or uncoupling vehicles
g. Installation of components onto a vehicle
h. Maintenance of a vehicle
i. Checking a vehicle
j. Installation and maintenance of permanent way
k. Inspection of track and structures

l. On-track machines (OTMS, including on-track plant) and permanent way maintenance vehicles for installation and maintenance
m. Signal engineering
n. Telecommunication systems
o. Supplying electricity to the transport system, to vehicles using it and to the telecommunications system
p. Controlling the supply of electricity to the electric traction and signalling systems
q. Communicating with signallers, electrical control operators,
r. Protecting persons on or near the track
s. Practical training and the supervision of practical training

5.4 Health Risk Management for Work Related to Safety

A risk management approach forms the basis for all health assessment decisions to ensure that the type of health assessment matches the risks associated with different works related to safety.

The health risk management process aims to:

- identify what could go wrong in the case of physical or psychological ill-health
- assess the consequences and
- establish appropriate controls for the risks associated with ill-health.

The health risk management process focuses on a consideration of the extent to which the staff's physical and/or psychological health could contribute to a serious incident on the rail network that may result in:

- the death of a person
- incapacitating injury to a person
- a collision or derailment involving rolling stock that results in significant damage or
- any other occurrence which results in significant property damage.

A further consideration is the extent to which the staff's health affects their own safety and that of fellow staff with a safety-critical position. Health assessments are one approach to treating the risk of serious incidents and the risk to individual safety; thus, a mix of engineering, administrative and health assessment measures are likely to be required.

Therefore, in determining the health assessment requirements of staff with a safety-critical position, it is important to take into account the operational and engineering environment, since overall risk management significantly determines the human attributes that are required for safety.

Health assessment standards must be set and applied carefully to match the risks associated with the tasks to be consistent with anti-discrimination and privacy laws.

5.5 Risk Categorization of Staff with a Position Critical to Safety

The health risk management approach is based on categories of risk, which help to define broad physical and psychological health attributes needed for particular tasks critical to safety. The approach also helps with the identification and monitoring of task-specific requirements.

The health assessment system is based on a risk analysis of work related to safety and categorisation of risk which is in turn based on considering the key question: *For any aspect of the tasks identified, could ill-health lead directly to a serious incident affecting the public or the rail network*?

This process is illustrated in Fig. 5.1. Two main risk categories are defined:

- Work critical to safety
- Safety non-critical Work

These two main categories are further divided, resulting in four risk categories overall (Category 1 to 4).

Staff with a safety-critical position
Staff with a safety-critical position are those whose action or inaction, due to ill-health, may lead directly to a serious incident affecting the public or the rail network. The staff's physical and psychological fitness to carry out their job is crucial. There are two risk categories for work critical to safety:

- High-Level Work critical to safety (Category 1)
- Work critical to safety (Category 2).

Tasks critical to safety are *High-Level safety-critical* if sudden staff incapacity such as a heart attack or blackout could result in a serious incident affecting the safety of the public or rail network. Single operator train driving on the mainline is an example of a High-Level safety-critical task (Category 1).

Safety-critical tasks which are not High Level include those where fail-safe mechanisms ensure sudden incapacity does not affect the safety of the rail network. For example, in many cases, a signalling task is safety-critical (Category 2) but not High-Level safety-critical because fail-safe systems ensure the safety of the network in case of staff incapacity.

To ensure sound physical and psychological health, the health assessments for staff with a safety-critical position (Category 1 and 2) require a comprehensive physical and psychological assessment. The assessments aim to detect conditions that may affect safe working, including heart disease, diabetes, epilepsy, sleep disorders, alcohol and drug dependence, psychiatric disorders and eye and ear problems.

Staff with a safety-critical position of Category 1 is required to undergo additional assessment for their risk of sudden incapacity. This involves a Cardiac Risk

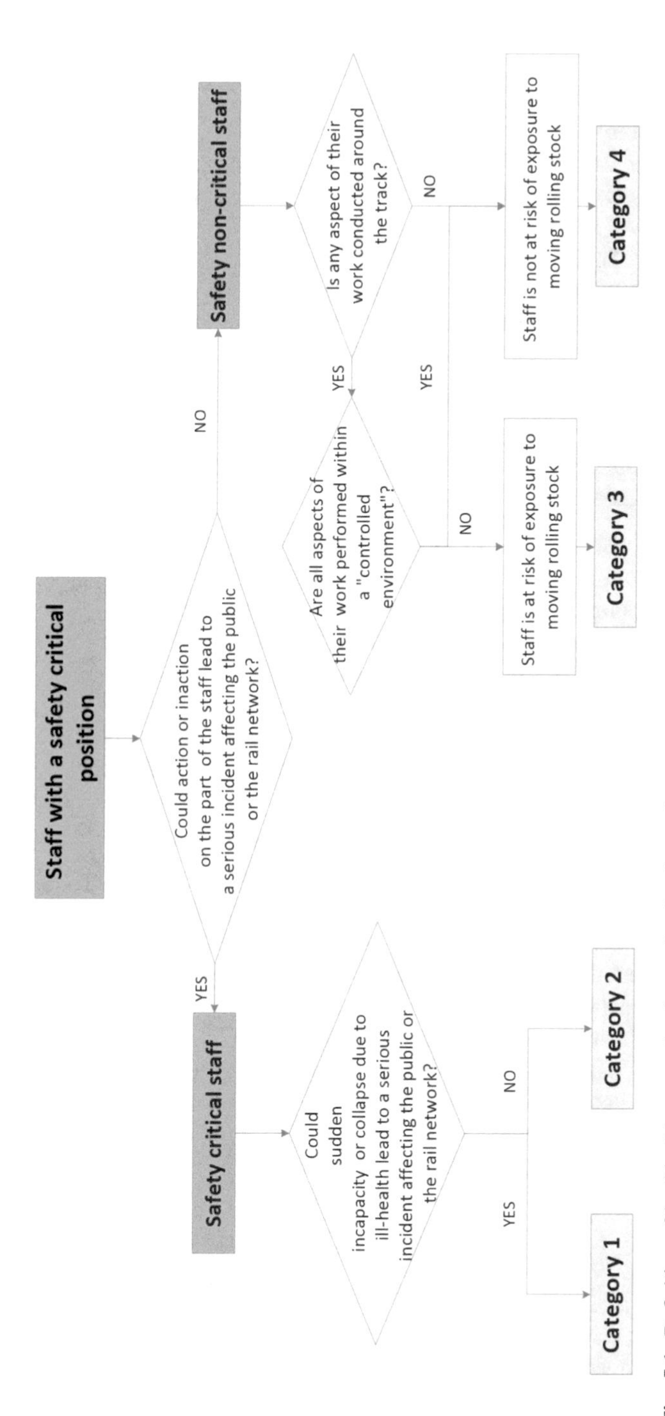

Fig. 5.1 Definition of health risk categories for work related to safety

Score assessment which is a screening tool to determine their risk of a cardiovascular event.

Staff with a Safety Non-Critical position

Staff with a Safety Non-Critical position are those whose health and fitness will not impact directly on the safety of the public and the rail network. This staff is categorised based on whether their health and fitness will impact on their ability to protect their own safety and that of fellow staff.

Around the Track Personnel is the term used to describe staff who perform Safety non-critical tasks on or near the track as defined. Their Risk Category depends on their likely exposure to moving rolling stock.

There are two staff risk categories with a Safety Non-Critical position:

- Around the Track Personnel operating in an uncontrolled environment (Category 3)
- Other, including Around the Track Personnel operating in a controlled environment[1] (Category 4).

Category 4 also includes those staff with a safety-critical position who do not work on or about the track, as illustrated in Fig. 5.1.

When analysing the risk to Around the Track Personnel and classifying the tasks into Categories 3 or 4, the features of a "controlled environment" need to be carefully considered regarding their adequacy.

Where Around the Track Personnel cannot be protected by a Controlled Environment, they must have the ability to sense an oncoming train and move quickly out of the way. They are therefore required to have health assessments commensurate with these risks, including hearing, vision and mobility (Category 3).

Note: staff directly working on the track should be regarded as functionally deaf and blind and require appropriate protection.

Staff in a Controlled Environment do not rely on their vision, hearing and mobility to protect them from risk and do not require a rail safety health assessment (Category 4), but they may require some form of assessment to meet the Health and Safety requirements of the job.

Where staff may move between controlled and uncontrolled environments, the higher level of risk assessment should be applied.

Irregular visitors to the track, such as office staff, are not generally classified as Around the Track Personnel. When they do visit the track, their safety should be ensured by other means, for example, by an escort.

---

[1]See definition (A Controlled Environment is defined as a rail workplace such as a siding, rail yard or workshop environment where a risk assessment has been performed to identify hazards and where controls are implemented to ensure that any person working in or transiting the area is not placed at risk from moving trains.)

5.6 The Process of Health Risk Management

This section presents the process of health risk management.

Rail Organizations should ensure that the process and rationale for the health assessment requirements of its staff with a safety-critical position are transparent. An effective risk management process will generally involve communication between the responsible manager and the staff who perform the safety-critical tasks.

The process should also rely upon appropriate expertise. Involvement of the Health Professional at the risk analysis stage will help identify necessary health attributes for a task. In turn, the Health Professional is likely to develop a sound understanding of the work and associated risks. When completing a risk assessment, it is important to clearly state the reasons why a task was so categorised. This may have legal significance in the future. The name of the person who made the assessment should be recorded.

The rail Organization should establish a procedure to ensure that the health risk management process and effectiveness of risk control strategies are kept under review. As a minimum, a review should occur whenever there are changes to work practices or engineering controls.

The process for health risk management is presented in Fig. 5.2 and described next in detail.

5.6.1 Step 1: The Context (Define the Context)

The first step is to establish and describe the context in which the process of managing health risks is to be carried out. This includes relevant legislative requirements, policies and procedures for organizations and the business and operational environment.

Rail Safety Standards

The documentation on the Rail Safety Standards concerned forms the basis for general risk management for safety-related work. It also determines the definition of staff with a safety-critical position and hence the scope of the health risk management process of an organization.

Business Environment

Identifying the business environment in which the organization operates also helps to establish the risk management framework.

For example:

- passenger train operations
- freight operations, including transport of dangerous goods
- infrastructure maintenance or construction.

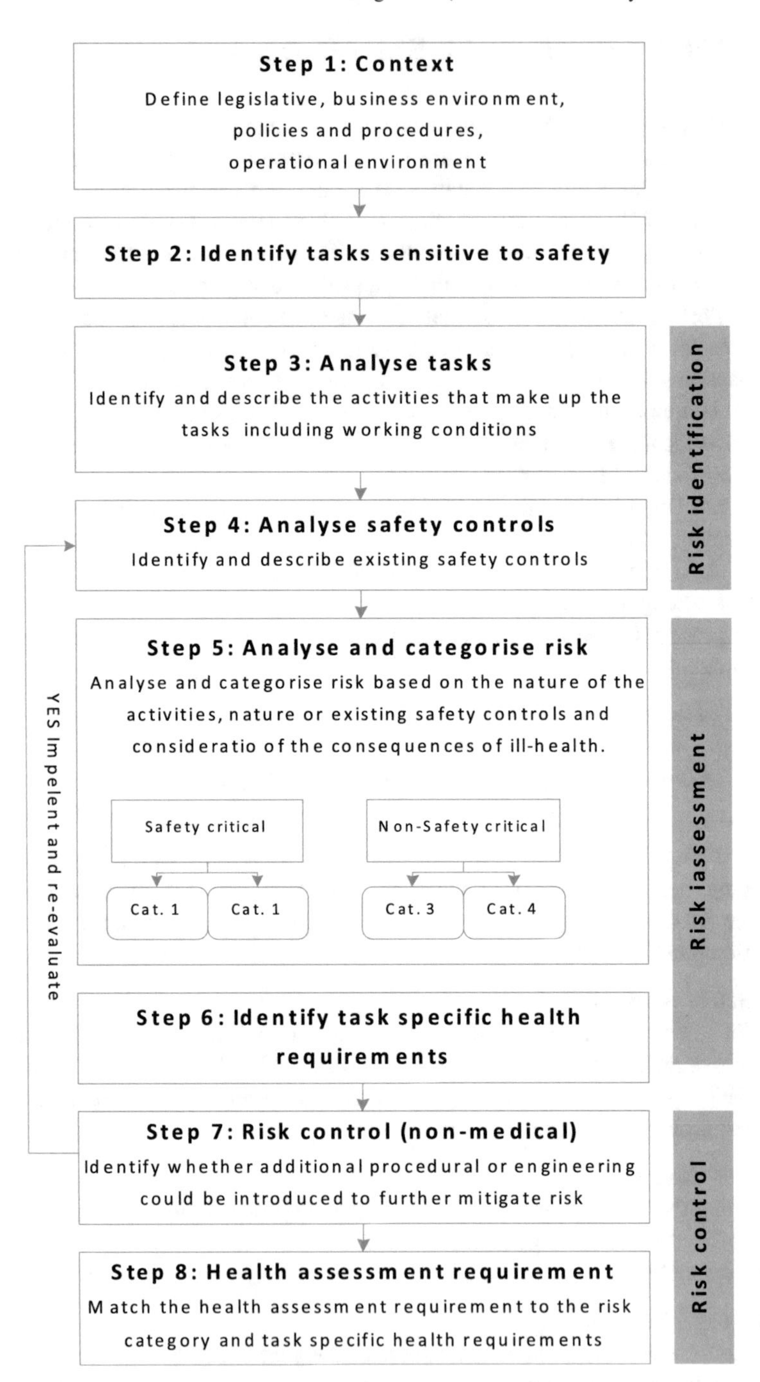

Fig. 5.2 Steps in the risk assessment process

Policies and Procedures
The process for managing health risks should be consistent with the rail Organization's general risk management framework. The policies and objectives of the organization for rail safety should be considered in order to help define the criteria for determining whether or not a risk is acceptable, and to determine the appropriate mix of engineering, administrative and medical control measures for risk management.

Operational Environment and Systems
An important contextual consideration is also the broad operational systems that support risk management in the Organization. Considerations may include:

- the type of safe-working systems
- train protection systems
- the maximum speed of operation.

5.6.2 Step 2: Identify Tasks Critical to Safety

After defining the broad context of health risk management, the next step is to identify and document all tasks performed within the Organization that are critical to the safety. These tasks will be the focus of the process of managing health risks which aims to:

- identify what could go wrong in the case of ill-health
- assess the consequences and
- establish appropriate controls for the risks associated with ill-health.

The analysis should focus initially on tasks and not on formal job classifications or grades. This is because the employees are often required to be multi-skilled and perform different tasks within one job.

Once tasks have been analysed, the analysis may then be applied to multi-skilled positions, with the highest risk task determining the level of health assessment required.

The following provides a list of typical jobs and tasks for a Rail Organization, which may include work critical to safety (see also Sect. 5.3).

(1) *Train driving*
(2) *Operation of signalling equipment*
(3) *Train controlling*
(4) *Infrastructure maintenance*
(5) *Rolling stock maintenance.*

5.6.3 Step 3: Analyse Tasks

After identifying all the safety-critical tasks, each task should be examined in order to determine the specific activities that make up that task.This may involve:

- A review of the job descriptions
- On-site visits to discuss tasks with safety-critical staff and to observe the tasks involved, as well as the conditions under which the activities are carried out and
- identifying activities infrequently performed in response to an emergency.

This step should also identify working conditions associated with the task, as these may also be relevant to the health requirements, for example:

- shift work
- working in extremes weather conditions or
- terrain etc.

5.6.4 Step 4: Analyse Safety Controls (Identify and Describe Local Safety Controls)

The nature of the operational and engineering environment will in part determine the human attributes that are required for safety. This includes the operational and/or engineering controls intended to mitigate the risk of the task. Therefore, the next step is to identify and evaluate the impact of local safety controls on the work being analysed on rail safety. For example:

- safe working rules and procedures
- fail-safe systems
- numbers of personnel in the working environment (such that other staff may identify staff incapacity and take up their task to ensure safety)
- driver support devices such as vigilance systems, train stops, the Automatic Warning System and Automatic Train Protection.

5.6.5 Step 5: Analyse and Categorise Tasks

The previous steps provide the inputs needed to categorize the staff with a critical safety position. This risk analysis is best conducted in conjunction with people who are knowledgeable of the tasks in question and the existing control measures.

In the analysis, the first consideration is whether the task is critical to safety or not. This is identified through the application of the test (refer to Sect. 5.5):

> For any aspect of the tasks identified, could action or inaction on the part of the staff lead directly to a serious incident affecting the public or the rail network?

This question arises in connection with existing control measures, such as vigilance systems and fail-safe mechanisms(as per Step 4). A further test is applied to subdivide the safety-critical tasks:

> For any aspect of the tasks identified, could sudden incapacity or collapse lead to a serious incident on the rail network?

Again, this question arises in the context of existing control measures and taking into account the likelihood of a serious incident resulting from incapacity on the part of the staff. The test leads to a subdivision of safety-critical tasks into Category 1 and Category 2 tasks as described in Sect. 5.5.

Example: Road-rail vehicle driver

> A road-rail vehicle has a sole driver, travels at up to 80 km/h and has a vigilance control (which brakes the vehicle if not regularly activated) but requires the driver to stop at level crossings. The task is considered safety-critical because the driver's continued vigilance is necessary to maintain appropriate control of the vehicle to ensure the safety of the rail network. In the event of sudden incapacity (e.g. a heart attack) just before a level crossing, the vehicle may enter the crossing before stopping. However, the likelihood of collapse occurring in the few hundred metres before a crossing is unlikely and therefore, the risk is analysed as low (Category 2). This contrasts with the driver of a track-tamper machine, which has a settable throttle, and without vigilance control, the collapse of a sole operator could lead to a large machine progressing out of control. Therefore, the risk is analysed as high (Category 1).

Categorising Safety Non-critical Work

Safety non-critical work is measured in a similar manner, resulting in classification to category 3 or category 4 depending on the assessment of the safety requirements of staff and fellow staff with a critical safety position and adequacy of measures to build a controlled environment. When analysing the risk to Around the Track Personnel and classifying the tasks into Categories 3 or 4, the method and adequacy of a Controlled Environment should be carefully considered in terms of its suitability.

Category 3 assessments relate to the ability of staff with a safety-critical position to see and move from the path of rail vehicles. In the case of a worksite where rail vehicles are being moved, a Category 3 assessment should be applied.

The determination of a Safety non-Critical Staff Category 4 depends on whether the work is performed in a **Controlled Environment**. When analysing the risk to Around the Track Personnel, the characteristics of a controlled environment must be defined and their appropriateness carefully considered. The essential requirement of a Controlled Environment is that it must ensure that a person transiting the area is not placed at risk from moving rolling stock, so far as reasonably practicable.

In rail workplaces, such as sidings, rail yards or workshops, controls may include:

- provision of lock-out or warning devices
- barrier segregation from running lines
- permits to work.

These may be supplemented as identified by risk assessment by all or any of the following:

- warning signage
- special instructions
- use of designated pathways or access/transit routes
- supervision.

For special works, a running line may also be assessed as a Controlled Environment in certain circumstances, for example, in the case of complete possession of all sections of track in the vicinity, including parallel lines.

5.6.6 Step 6: Identify Task-Specific Health Requirements

Some health requirements are independent of the risk category. These include sensory requirements, such as hearing and colour vision, as well as musculoskeletal requirements. Rail Organizations should conduct risk assessments of individual tasks to identify the requirements. These requirements should be communicated to Health Professionals when requesting a health assessment.

Colour vision risk assessment Assessment of a job requires consideration of whether there is a need for colour vision. Not all tasks critical to safety require colour vision; thus, risk assessments of the colour vision requirements should be undertaken by the Rail Organization and communicated to the Health Professional.

Hearing risk assessment The hearing requirements vary for different tasks and are generally independent of the overall risk category (except for Category 3). For example, a train driver must be able to communicate with control about train orders, often in a noisy cab. This requires sufficient hearing to interpret speech accurately. Alternatively, a track staff only requires sufficient hearing to detect the sound of a train horn or warning shouts from other staff.

All safety-critical tasks should be assessed in relation to their individual hearing requirements.

Musculoskeletal requirements The standard [1] for both Category 1 and Category 2 staff requires the staff to be fit enough to undertake the physical demands of their safety-critical position. In the case of Category 3 staff, the assessment focuses on their mobility and capacity to move quickly from the path of an oncoming train. This should cover most situations in rail work, but the health assessment may be varied depending on the result of the task evaluation and on expert medical advice. For example, a controller may not need lower limb function, whereas a rolling stock maintainer requires considerable agility to move and inspect trains.

Table 5.1 Summary of the hierarchy of control measures

Elimination	Permanent solution should be attempted in the first instance.
Isolation	Replace hazard or environmental aspects by one of lower risk.
Minimisation	Lowest level of control.

5.6.7 Step 7: Risk Control

The categorisation of health risks undertaken in Phase 6 is the basis of a balanced health assessment . However, an important interim step is to consider the other treatment options that might be introduced to mitigate the risk, such as additional administrative or engineering controls.

Table 5.1 summarises the hierarchy of control measures that may be applied to control safety risks.

Both elimination and isolation control the hazard itself. They are, therefore, more effective in reducing risk than controls which reduce the likelihood of the hazard.

If practicable, minimisation controls are generally preferred to health assessments as they provide more definitive protection. Such improvements should be implemented where possible and the task re-evaluated in terms of health risk.

5.6.8 Step 8: Confirm Health Assessment Requirements

After health assessment determining the final risk categories of staff with a safety-critical position, the health assessments are matched to the categories, i.e. Category 1 are required to have a "High-Level critical Staff" health assessment; Category 2 is required to have a "critical staff position" assessment.

Occupational Health and Safety (OHS)
Staff may also be required to have appropriate OHS examinations for specific hazards such as noise, asbestos or hazardous substances.

Reference

1. Australia, National Transport Commission (2012) National Standard for Health Assessment of Staff with a safety-critical position, October 2012 as amended up to 16 March 2013

Definitions

Acceptable risk	Level of risk that is accepted in a given context based on the current values of society (the terms "acceptable risk" and "tolerable risk" are considered to be synonymous) (ISO/IEC guide 51:2014)
Actor	Any party which is, directly or through contractual arrangements, involved in the application of the risk management process.
Accident	An unwanted or unintended sudden event or a specific chain of such events which have harmful consequences; accidents are divided into the following categories: collisions, derailments, level-crossing accidents, accidents to persons caused by rolling stock in motion, fires and others (EU Directive 2001/14/EC)
Assessment body	An independent and competent person, Organization or entity which undertakes investigation to arrive at a judgment, based on evidence, of the suitability of a system to fulfil its safety requirements
Assessment report	The document containing the conclusions of the assessment performed by an assessment body on the system under assessment
Communication and Consultation	Continual and iterative processes that an Organization conducts to provide, share or obtain information and to engage in dialogue with Stakeholders regarding the management of risk
Change	Any change to an Organization's procedure or operation that significantly alters the risk profile of the Organization. Change is anything that alters or modifies present practice
Consequence	Outcome or impact of an event (There can be more than one consequence from one event Consequences can range from positive to negative Consequences can be expressed qualitatively or quantitatively Consequences are considered in relation to the achievement of objectives.)

(continued)

K. Tzanakakis, *Managing Risks in the Railway System*, Springer Tracts on Transportation and Traffic 18, https://doi.org/10.1007/978-3-030-66266-0

(continued)

Control	A measure that is modifying risk. An existing process, policy, device, practice or other action that acts to minimize negative risk or enhance positive opportunities Note: the word 'control' may also be applied to a process designed to provide reasonable assurance regarding the achievement of objectives
Control assessment	A systematic review of processes to ensure that controls are still effective and appropriate Note: periodic line management review of controls is often called 'control self-assessment'
Establishing the context	Defining the external and internal parameters to be taken into account when managing risk, and setting the scope and risk criteria for the risk management policy
Event	Occurrence or change of a particular set of circumstances Note 1: the event can be certain or uncertain Note 2: the event can be a single occurrence or a series of occurrences (ISO/IEC guide 73, in part)
External context	The external environment in which the Organization seeks to achieve its objectives
Frequency	A measure of the number of occurrences per unit of time
Hazard	• A system condition that could lead to an accident. A source of potential harm (ISO/IEC Guide 51) • An activity, arrangement, circumstance, event, occurrence, phenomenon, process, situation, or substance (whether arising or caused within or outside a place of work) that is an actual or potential cause or source of harm
Hazard record	The document in which identified hazards, their related measures, their origin and the reference to the Organization which has to manage them are recorded and referenced
Internal context	Internal environment in which the Organization seeks to achieve its objectives
Level of Risk	The magnitude of risk or combination of risks expressed in terms of the combination of consequences and their Likelihood
Likelihood	Chance of something happening. It is used as a general description of probability or frequency
Loss	Any negative consequence or adverse effect, financial or otherwise
Monitor	To check, supervise, observe critically or measure the progress of an activity, action or system on a regular basis in order to identify change from the performance level required or expected
Probability	A measure of the chance of occurrence expressed as a number between 0 and 1
Proposer	A person making the change (technical, operational or Organizational change)
Railway system	The totality of the subsystems for structural and operational areas, as defined in Directives 96/48/EC and 2001/16/EC, as well as the management and operation of the system as a whole

(continued)

(continued)

Residual risk	Risk remaining after implementation of risk treatment
Review	Activity undertaken to determine the suitability, adequacy and effectiveness of the subject matter to achieve established objectives
Risk	The frequency of occurrence of accidents and incidents resulting in harm (caused by a hazard) and the degree of severity of that harm
Risk analysis	The systematic use of all available information to identify hazards and to estimate the risk.
Risk appetite	Level of risk that an Organization can undertake and successfully manage over an extended time period
Risk assessment	The overall process of risk identification, risk analysis and risk evaluation
Risk attitude	Organization's approach to assess and eventually pursue, retain, take or turn away from risk
Risk avoidance	A decision not to become involved in, or to withdraw from, a risk situation
Risk criteria	Terms of reference by which the significance of risk is assessed Note: risk criteria can include associated cost and benefits, legal and statutory requirements, socioeconomic and environmental aspects, the concerns of stakeholders, priorities and other inputs to the assessment
Risk evaluation	• A procedure based on the risk analysis to determine whether the acceptable risk has been achieved. It is the process of comparing the results of risk analysis with risk criteria to determine whether the risk and/or its magnitude is acceptable or tolerable • A procedure based on the risk analysis to determine whether tolerable risk has been exceeded (ISO/IEC GUIDE 51:2014)
Risk identification	Process of finding, recognizing and describing risks
Risk identification	Process of finding, recognizing and describing risks. It is the process of determining what, where, when, why and how something could happen
Risk management	• Coordinated activities to direct and control an Organization with regard to risk • The culture, processes and structures that are directed towards realizing potential opportunities whilst managing adverse effects
Risk management framework	Set of components that provide the foundations and organizational arrangements for designing, implementing, monitoring, reviewing and continually improving risk management throughout the Organization
Risk management plan	Scheme within the risk management framework specifying the approach, the management components and resources to be applied to the management of risk
Risk management policy	Statement of the overall intentions and direction of an Organization related to risk management
Risk management process	Systematic application of management policies, procedures and practices to the activities of communicating, consulting, establishing the Context, and identifying, analysing, evaluating, treating, monitoring and reviewing risk

(continued)

(continued)

Risk owner	Person or entity with the accountability and authority to manage a risk
Risk management philosophy	The general attitude or approach an Organization takes in dealing with risks
Risk profile	Description of any set of risks
Risk reduction	Actions taken to lessen the likelihood, negative Consequences, or both, associated with a risk
Risk retention	Acceptance of the burden of loss, or benefit of gain, from a particular risk (risk retention includes the acceptance of risks that have not been identified The level of risk retained may depend on risk criteria.) (ISO/IEC guide 73, in part)
Risk sharing	Sharing with another party the burden of loss, or benefit of gain from a particular risk (legal or statutory requirements can limit, prohibit or mandate the sharing of some risks. Risk sharing can be carried out through insurance or other agreements. Risk sharing can create new risks or modify an existing risk.)
Risk source	An element which alone or in combination has the intrinsic potential to give rise to risk
Risk treatment	Process of selection and implementation of measures to modify Risk (the term 'risk treatment' is sometimes used for the measures themselves. Risk treatment measures can include avoiding, modifying, sharing or retaining risk.) (ISO/IEC guide 73, in part)
Safety	Freedom from risk which is not tolerable (ISO/IEC GUIDE 51:2014)
Safety measure	As defined in the regulation: "A set of actions that either reduce the rate of occurrence of a hazard or mitigate its consequences in order to achieve and / or maintain an acceptable level of risk." (Article 3, clause10)
Safety requirement	As used in this guidance: A characteristic of a system and its operation (including operational rules) necessary in order to deliver acceptable risk
Serious accident	Any train collision or derailment of trains, resulting in the death of at least one person or serious injuries to five or more persons or extensive damage to rolling stock, the infrastructure or the environment, and any other similar accident with an obvious impact on railway safety regulation or the management of safety; 'extensive damage' means damage that can immediately be assessed by the investigating body to cost at least EUR 2 million in total (EU Directive 2001/14/EC)

(continued)

(continued)

Stakeholder	Person or Organization that can affect, be affected by, or perceive themselves to be affected by a decision or activity
System	That part of the railway system which is subject to a change
Tolerable risk	Level of risk that is accepted in a given context based on the current values of society (the terms “acceptable risk” and “tolerable risk” are considered to be synonymous)

Appendix A: Definition of Fatality and Weighted Injury (FWI) as Defined in the British Rail Industry

(Based on RSSB (Rail Safety and Standards Board), (UK), "*Guidance on Hazard Identification and Classification*", Rail Industry Guidance Note, June 2014)

In order to quantify, compare and understand the impact of safety-related incidents and risk the Great Britain railway industry has adopted, by industry agreement, the scheme set out in the table below for relative weighting of different types of injury (Table A.1).

K. Tzanakakis, *Managing Risks in the Railway System*, Springer Tracts on Transportation and Traffic 18, https://doi.org/10.1007/978-3-030-66266-0

Table A.1 Definition of fatality and weighted injury

Injury degree	Definition	Weighting	Ratio
Fatality	Death occurs within one year of the accident	1	1
Major injury	Injuries to passengers, staff or members of the public as defined in schedule 1 to RIDDOR 1995. This includes losing consciousness, most fractures, major dislocations, loss of sight (temporary or permanent) and other injuries that resulted in hospital attendance for more than 24 h	0.1	10
Class 1 minor injury	Injuries to passengers, staff or members of the public, that are neither fatalities nor major injuries, and are defined as reportable in RIDDOR 1995[1] amended April 2012, and workforce injuries, where the injured person is incapacitated for their normal duties for more than three consecutive calendar days, not including the day of the injury	0.005	200
Class 2 minor injury	All other physical injuries	0.001	1000
Class 1 shock/trauma	Shock or trauma resulting from being involved in, or witnessing, events that have serious potential of a fatal outcome, for example train accidents such as collisions and derailments, or a person being struck by train	0.005	200
Class 2 shock/trauma	Shock or trauma resulting from other causes, such as verbal abuse and near misses, or personal accidents of a typically non-fatal outcome	0.001	1000

Appendix B: The Railway System and Its Subsystems

The railway system comprises of subsystems. Each subsystem provides a set of functionalities, which, combined with the other subsystems, provide integrated functionality (Fig. B.1).

The high-level subsystems are categorised as the three different types of approvals, as per EN50129:2003:

(1) A "Generic Product (GP)" approval (the platform).
(2) A "Generic Application (GA)" approval (the type).
(3) A "Specific Application (SA)" approval (the installed product).

Approval type		
Specific application (SA)	Generic application (GA)	Generic product (GP)
A. Infrastructure	A1. Railway Track	• Rails • Sleepers • Ballast • Switches and crossings • Special track forms in depot areas
	A2. Civil work	• Embankments • Cuts • Bridges • Tunnels • Station buildings

(continued)

K. Tzanakakis, *Managing Risks in the Railway System*, Springer Tracts on Transportation and Traffic 18, https://doi.org/10.1007/978-3-030-66266-0

(continued)

Approval type		
Specific application (SA)	Generic application (GA)	Generic product (GP)
		• Other civil work (underpasses, overpasses, animal crossings, drainage systems, noise barriers, fencing)
B. Technological systems	B1. Signalling and train control system	Main subsystems involved, and their components are listed below: • Centralised traffic control • Computer-based interlocking • Automatic train control and train protection • Train detection • Trackside devices • Signalling technical buildings • Signalling power Supply[1] • Auxiliary defect detectors
	B2. Telecommunication systems	Communications systems include: • Multi-Service Network (CCNN) and Fibre Optic Network (FON) • Local Area Networks • Telephone system • Help points • Voice recording • Public address system • Master Clock • CCTV/Intruder Detection/Access Control System
	B3. Telematics applications	This subsystem comprises two elements: a. applications for passenger services, including systems which provide passengers with information before and during the journey, reservation and payment systems, luggage management and management of connections between trains and with other modes of transport; b. applications for freight services, including information systems (real-time monitoring of freight and trains), marshalling and allocation systems, reservation, payment and invoicing systems, management of connections with other modes of transport and production of accompanying electronic documents.

(continued)

[1]Track-side equipment, and in general, every vital signalling device will be fed by no-break power supply systems in order to achieve a continuous service availability of the safety apparatus.

(continued)

Approval type		
Specific application (SA)	Generic application (GA)	Generic product (GP)
C. Safety and security systems:	C1. Fire detection systems C2. Fire Fighting systems C3. Emergency lightings C4. Access control systems C5. Intruder detection systems C6. CCTV	
D. Ancillary systems	D1. Mechanical systems D2. Electrical systems D3. Plumbing systems	Mechanical services comprise of the following equipment: • Heating • Air conditioning (HVAC) • Ventilation Electrical services comprise of the following equipment: • Low voltage (LV) distribution systems • Complete fit-out of switchboards, switchgear and equipment • Cable containment for electrical LV services • Cable pits and conduit system for external LV services • Local control system • Earthing and bonding • Mobile generators • Normal lighting Public health services (Plumbing) are provided in the following areas: • Passenger stations • Freight facilities • Wayside buildings • Depots
E. Energy	The electrification system, including overhead lines and the trackside electricity consumption measuring and charging system	
F. Rolling stock		
G. Maintenance	Logistic and Maintenance infrastructure (Marshalling Yards, Maintenance Depots, Container Terminals, Wayside Maintenance Buildings) (The procedures, associated equipment, logistics centres for maintenance work and reserves providing the mandatory corrective and preventive maintenance to guarantee the performance required)	
H. Operation and traffic management	The procedures and related equipment enabling a coherent operation of the different technical subsystems, both during normal and degraded operation, including train driving, traffic planning and management	

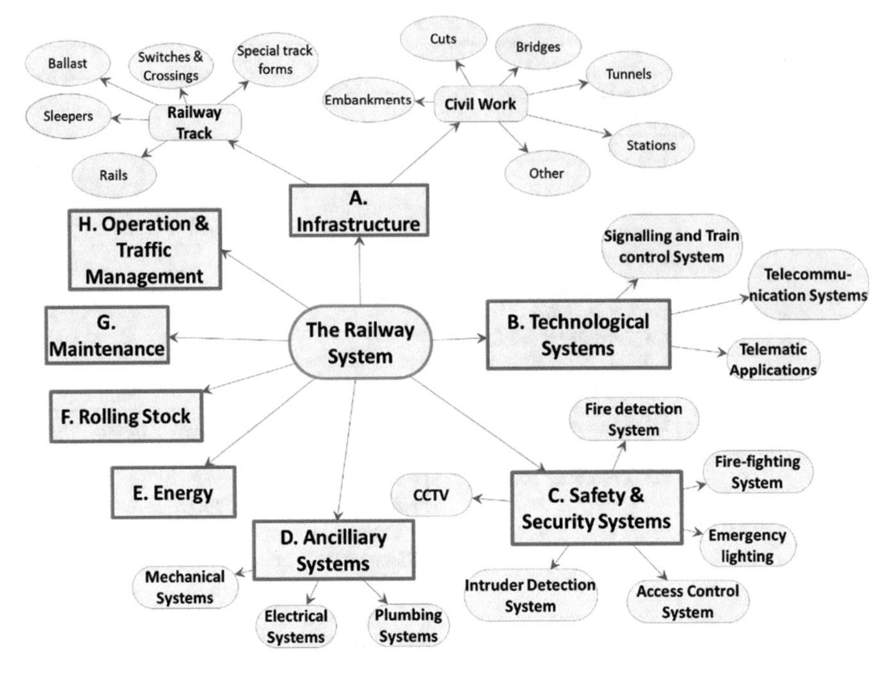

Fig. B.1 The railway system and its subsystems and component examples

Appendix C: Laws, Codes and Standards

BS 6079-3:2000	Project Management—Part 3: Guide to the Management of Business-related Project Risk (British Standards Institution)
BS IEC 61882:2001	Hazard and operability studies (HAZOP studies). Application guide
CAN/CSA-Q850-97	Risk Management: Guideline for Decision Makers (Canadian Standards Association)
CENELEC, "CLC/TR 50126-2	Guide to the application of EN 50126-1 for safety", 2009
CENELEC, "EN 50126:1999	Railway Applications. The Specification and Demonstration of Reliability, Availability, Maintainability and Safety (RAMS). Generic RAMS Process"
CLC/TR 50126-2:2007	Railway applications—the specification and demonstration of reliability, RAMs
EC No 352/2009	Commission Regulation on a Common Safety Method on risk evaluation and assessment
EN 50126-1:1999	Railway applications—The specification and demonstration of Reliability, Availability, Maintainability and Safety (RAMS)
EN 60812:2006	Analysis techniques for system reliability. Procedure for failure mode and effects analysis (FMEA)
EN 60812:2006	Analysis techniques for system reliability. Procedure for failure mode and effects analysis (FMEA)
EN 61025:2007	Fault tree analysis (FTA))
EN 61508-1:2010	Functional safety of electrical/electronic/programmable electronic safety-related systems. General requirements
EN 61882:2016	Hazard and operability studies (HAZOP studies)
EN 62502:2011	Analysis techniques for dependability. Event tree analysis (ETA)
EU No 402/2013	Commission Implementing Regulation on a Common
IEC 62198:2001	Project Risk Management—Application Guidelines (International Electrotechnical Commission)
IRM/Alarm/AIRMIC 2002	A Risk Management Standard (Institute of Risk Management (IRM), Association of Insurance and Risk Managers (AIRMIC) and National

(continued)

K. Tzanakakis, *Managing Risks in the Railway System*, Springer Tracts on Transportation and Traffic 18, https://doi.org/10.1007/978-3-030-66266-0

(continued)

	Forum for Risk Management in the Public Sector (ALARM), London, UK)
ISO 14004	Environmental management systems—General guidelines on principals, systems and supporting techniques
ISO 14050	Environmental management—Vocabulary
ISO 15489	Records management
ISO 31000 / 2009	Risk Management—Principles and Guideline
ISO 3534-1	Statistics; Vocabulary and symbols; Part 1: Probability and general statistical terms
ISO 9000	Quality management systems—Fundamentals and vocabulary
ISO/IEC 31010 / 2009	Risk Management—Risk Assessment Techniques
ISO/IEC Guide 51	Safety aspects—Guidelines for their inclusion in standards
ISO/IEC Guide 73	Risk Management—Vocabulary—Guideline for use in standards
ISO/IEC Guide 73	Risk management—Vocabulary— Guidelines for use in standards
ISO/TR 31004 / 2013	Risk Management—Guidance for the Implementation of ISO 31000
JIS Q 2001:2001(E)	Guidelines for Development and Implementation of Risk Management System (Japanese Standards Association)
PD 6668:2000	Managing Risk for Corporate Governance (British Standards Institution)

Other Documents

1. ERA/GUI/02-2008/SAF European Railway Agency Collection of examples of risk assessments and of some possible tools supporting the CSM Regulation
2. RSSB, GE/GN8640, Guidance on Planning an Application of a CSM on Risk Evaluation and Assessment
3. RSSB, GE/GN8641, Guidance on System Definition
4. RSSB, GE/GN8643, Guidance on Risk Evaluation and Risk Acceptance
5. RSSB, GE/GN8644, Guidance on Safety Requirements and Hazard Management
6. RSSB, GE/GN8645, Guidance on Independent Assessment
7. RSSB, GD-0001-SKP, Taking Safe Decisions—how Britain's railways take decisions that affect safety
8. RSSB, Research project T270 Railway Action Reliability Assessment: A technique for quantification of human error in the rail industry
9. RSSB, Research project T440, The weighting of non-fatal injuries: Fatalities and weighted
10. RSSB, Injuries Research project T955 Hazard analysis and risk assessment for rail projects
11. ORR, Safety Method for risk evaluation and assessment (Dec 2012 ORR guidance on the application of the common safety method (CSM) on risk assessment and evaluation (December 2012)

12. RIDDOR—Reporting of Injuries, Diseases and Dangerous Occurrences Regulations, 1995)
13. Australian/New Zealand Standard, *"AS/NZS 4360 SET Risk Management"*, 2004
14. New Zealand, "*National Rail System Standard/4— Risk Management*", 2007
15. Canadian Standards Association, "*Risk Management: Guideline for Decision-Makers (CAN/CSA-Q850-97)*", 1997, Reaffirmed 2002
16. Standards Australia / Standards New Zealand, "*Risk Management Guidelines—Companion to AS/NZS 4360:2004*", 2005
17. Intergovernmental Organization for International Carriage by Rail (OTIF), "*Uniform Technical Prescription—Common Safety Method on risk evaluation and assessment (UTP GEN-G consolidated version)*", 1.12.2016

Appendix D: Professional Bodies Dealing with Risk Management

Next an indicative list of professional bodies, dealing with Risk Management is provided.

Association of Insurance and Risk (AIRMIC)	http://www.airmic.com
European Institute of Risk Management (EIRM)	http://www.eirm.com
European Union Agency for Railways (ERA)	https://www.era.europa.eu/
Federation of European Risk Management Associations	http://www.ferma-asso.org
Global Association of Risk Professionals (GARP)	https://www.garp.org/
Institute of Risk Management (IRM)	http://www.theirm.org
Pan-Asia Risk and Insurance Management Association (PARIMA)	http://parima.org/
Professional Risk Managers' International Association	http://prmia.org
Public Risk Management Association	https://primacentral.org/
Rail Safety Standards Board (RSSB) (UK)	https://www.rssb.co.uk/
Risk Management Association	https://www.rmahq.org/
Risk Management Association Australia	https://www.rmahq.org/
Risk Management Association of India (RMAI)	rmaindia.org/
Risk Management Institution of Australasia	http://www.rmia.org.au
Society for Risk Analysis (SRA)	http://www.sra.org
Society of Risk Management Consultants	https://srmcsociety.org/
The Institute of Risk Management South Africa	https://www.irmsa.org.za/

(continued)

K. Tzanakakis, *Managing Risks in the Railway System*, Springer Tracts on Transportation and Traffic 18, https://doi.org/10.1007/978-3-030-66266-0

(continued)

The International Federation of Risk and Insurance Management Associations (IFRIMA)	www.ifrima.org/
The Risk Management Society (RIMS)	www.rims.org
UNESCO, Bureau of Strategic Planning	http://www.unesco.org/new/en/bureau-of-strategic-planning/themes/risk-management/
University Risk Management and Insurance Association (URMIA)	https://www.urmia.org/

Index

K. Tzanakakis, *Managing Risks in the Railway System*, Springer Tracts on Transportation and Traffic 18, https://doi.org/10.1007/978-3-030-66266-0

Printed in the United States
by Baker & Taylor Publisher Services